AF323461

Phytopathological Classics

NUMBER 12

PREFACE

The following translation is a labor of love, three years in birth. It was our hope to have the translation in print by 1974 - the one hundredth anniversary of the original paper, and the one hundredth anniversary of the beginning of scientific forest and wood products pathology. Due to the pressures of other committments, we have missed this deadline somewhat, but are sufficiently close that this can well serve to celebrate the founding of these important branches of the science of phytopathology.

We started to make a free translation of this fundamental paper. But we became so completely enthralled by the flow of Hartig's prose that we decided to stay as close to a verbatim translation as possible. Paragraph and sentence length have been unchanged, but we freely applied commas, semicolons, colons, and dashes to break some of the rather long sentences into digestible phrases. In some instances we rearranged the German sentence structure, changed tenses of verbs and singular and plurals to make the sentence structure and grammar conform more closely to contemporary English usage. In all such cases, we have double checked to be sure that the sense of what Hartig was saying was not changed.

William Merrill
David H. Lambert

University Park, Pa.

Walter Liese

Hamburg, Germany

October 15, 1974

(Plates have been reduced to 9/10 of original size.)

Robert Hartig (1839-1901)

ROBERT HARTIG

1839 - 1901

Robert Hartig was born May 30, 1839, in Braunschweig, Germany, the third and last in a line of famous foresters. His grandfather, Georg Ludwig Hartig, was chief forester of Prussia and laid the foundations of scientific silviculture in Germany. In 1836, he also carried out field tests on the behavior of wood species towards fungi. Robert Hartig's father, Theodor Hartig (1805-1880), his outspoken idol, was Professor of Forestry at Braunschweig, and may be regarded as the first forest botanist. He contributed much to the sciences of forestry and botany, including his theory of forest yield and his discovery of sieve elements, callose, and aleuron grains.

Growing up in such a tradition and atmosphere, it was Robert Hartig's aim from childhood to follow in his father's and grandfather's footsteps. He began his studies in forestry at Braunschweig in 1861, entered the University of Berlin in 1863, and later joined the Forest Service of Braunschweig. His university studies together with family influences gave him a universal training with a broad foundation of general knowledge.

During his first professional years, he accumulated many observations on tree growth and yield in various parts of Germany; these formed the basis for his first book, published in 1865, on the growth of beech and oak under different conditions. However, daily forestry life with its narrow-minded administrative work annoyed him, so he began further formal studies and earned the doctoral degree in 1866 at Marburg University. In 1867 he was requested to join the Forestry Academy Eberswalde, where he first taught botany and zoology as a substitute for the ailing Professor Ratzeburg. In 1871 he became Professor of Botany.

The twelve years spent at Eberswalde determined his further career. At this time some knowledge was available about the attack of trees by insects, but almost nothing was known about the biology and importance of tree diseases caused by fungi. The then existent contributions of de Bary, Tulasne, and Kühn on the life cycles and hosts of microorganisms were related only to agronomic crops. Hartig concentrated on the fungi attacking the foliage, stems, and roots of the tree species around Eberswalde, and in 1874 produced his first masterpiece, "Wichtige Krankheiten der Waldbäume." This 124 page book clearly written and furnished with his own accurate illustrations of his astonishingly detailed observations, contains, besides a general introduction into the morphology and physiology of fungi, monographic treatments on *Agaricus melleus (Armillaria mellea), Trametes (Fomes) pini, Trametes radiciperda (Fomes annosus), Peridermium pini (Cronartium flaccidum),*

Caeoma pinitorquum (Melampsora pinitorqua), C. laricis (Melampsora spp.), Peziza (Trichoscyphella) Willkommii, and others.

This work must be examined in the context of the state of knowledge at that time. The germ theory of disease in plants had been recently proven by de Bary and others, but was still controversial. The germ theory of disease in animals by Koch had yet to be proven. There was still belief in spontaneous generation and a widespread disbelief—especially among practicing foresters—that fungi cause disease.

Georg Hartig had regarded decay as a natural process of metamorphosis during the aging of wood. Theodor Hartig (1833) had connected the occurrence of brown rot and white rot with the presence of fungal hyphae, but did not discover the cause and effect relationship. He still believed that decay was brought about by the age of the tree and the functionlessness of its organs. He postulated that during decay the wood cell structure broke up into little balls or monads which then formed into rows and fused with one another to form fungus hyphae.

H.v. Schacht (1863) described in detail the effects of fungal hyphae on wood of *Dracaena draco.* He described the symptoms of soft-rot of wood and emphasized the importance of moisture and aeration for fungal activity. However, he concluded, ". . . fungi, *feeding on decay products,* penetrate into the interiors of dead or rotting plant parts and *cause by themselves all sorts of destruction.* They are, indeed, the never absent companions of the decay phenomenon . . ." Thus, he too missed the cause and effect relationship, and did not understand that the softening and breakdown of the wood were merely companion effects of the fungi which he observed digesting bore holes through the cell walls.

Willkomm (1866) argued with Theodor Hartig about the spontaneous generation in which the latter believed, but which Willkomm rejected. In his investigations of the cause of the red rot (Rotfäule) of spruce, he clearly stated that a parasitic fungus (*Xenodrochus ligniperda*) must invade the stem and alter the wood by chemical deterioration. However, he believed in the origin of the mycelium from spores of other fungi and although some of his observations were correct and his ideas astute, he had no convincing proof.

Robert Hartig concluded that wood decay is caused by microorganisms, themselves independent organisms, and that each of the various patterns of decay is caused by a different agent. These conclusions were based on detailed observations and confirmed without doubt by artificial inoculations. Likewise, he provided the proof of pathogenicity of several fungi attacking other portions of the tree. This was the cornerstone for all further forest pathology, and well earned him the title, FATHER OF FOREST PATHOLOGY.

Only four years after his book of 1874, another fundamental contribution emerged from his pen: "Die Zersetzungserscheinungen des Holzes der Nadelbäume und der Eiche." This 151 page book is even more basic in nature, and is still used as a reference point for several of the decay fungi. He associated both macroscopic and microscopic symptoms with various fungi, and attributed these symptoms to specific enzymes. These two works laid the foundations for the whole field of wood products pathology, brought about fundamental changes in concepts in the field of wood preservation, led to much of the field of biodeterioration, and are in actuality the first basic treatises of pathological anatomy. The emphasis on pathological anatomy occurs throughout his writings and is emphasized in his texts.

Hartig was also interested in other fields of forest botany; he was the first to associate the sieve tube elements, discovered by his father, with phloem transport, through which the assimilate is transported to all parts of the vascular plant (1873). Besides his teaching obligations and scientific activity, he worked also in the city council and did much for the welfare of the community. His complete satisfaction in his work at Eberswalde is expressively shown by his refusal of an offer from the Forest Academy Aschaffenburg. But ultimately he could not resist taking up the post of Professor of Forest Botany at the famous University of Munich and Head of the Botanical Department of the Bavarian Forest Experimental Station in 1878.

In Munich he continued his studies on tree diseases, but developed also a strong interest in the anatomy and physiology of forest trees. His Chair at the Forestry Faculty still carries today the title, "Anatomy, Physiology, and Pathology of Plants," created by him. Until his death a great number of papers and books appeared, covering a wide range of subjects: forest pathology; wood products pathology; forest entomology; cambial activity; distribution of water, gas, and organic substances within a tree; influence of pollution on trees; water uptake by roots in winter; mechanics of sap movement; formation of the annual ring with early wood and late wood; wood quality; wood identification; cell wall structure; function of bordered pits; lightning effects on trees; spiral grain; compression wood; biology of *Merulius lacrymans* and *Polyporus vaporarius*. Besides the fungi already mentioned, he investigated for the first time several others, including *Phytophthora omnivora, Rosellinia quercina,* and *Dematophora necatrix.*

Looking at his list of publications and considering their quantity and quality, one can only feel highest admiration for such a fertile mind. In fact, there is hardly any field in classical forest botany and forest pathology which was not founded or further elaborated by Robert Hartig. He was a master in observing nature in its manifold forms and in capturing the typical characters in excellent drawings. His observations are so accurate and of such a high standard that one can benefit from them even today with all our sophisticated state of knowledge acquisition. In his papers he always stressed the close connection between structure and function of the tissue, organs, and individuals. Hartig successfully combined in his work pure and applied sciences at their best. Many of his findings soon became common knowledge. Obviously Hartig's work was not free from errors and some of his results were criticized; but mostly he was in the right.

The first textbook in forest pathology, "Lehrbuch der Baumkrankheiten" appeared in 1882, and in second and third editions in 1889 and 1900, respectively, as "Lehrbuch der Pflanzenkrankheiten." It was translated into English, French, and Russian and used for years throughout the world; it is still a basic reference. His text on Forest Botany published in 1891 was an outstanding compilation of the knowledge at that time about structure, growth and physiology of forest trees.

Robert Hartig also contributed much to the scientific foundation of silviculture. During his time, German State Forestry Research became organized and he fought powerfully for freedom in research combined with full responsibility of the individual scientist. He preferred to work on his own, as was characteristic of the great scientists of those days. He was a hard worker and an optimist; his cheerful personality was combined with discipline and humor. His lectures, well prepared, stimulating and genuinely given, were always accompanied by demonstrations; people from far away came to listen to him.

If one considers just the impact of Hartig's work on forest pathology, one can only admire the magnitude and depth of his contributions. After reading his descriptions of certain diseases, like those following, one has to admit that relatively little has been added to our knowledge on symptomology and mode of action during the past 100 years!

Hartig's publications—about 170 in number—were published in various journals. Therefore, he assembled results of many previous studies together with results of more recent studies in his book, "Holzuntersuchungen, Altes and Neues," which appeared only after his death. His studies were continued by his successor and son-in-law, Professor Dr. Carl Freiherr von Tubeuf, who himself made substantial contributions to forest pathology, including development of isolation techniques and procedures for laboratory work with pure cultures.

Hartig was ill throughout his later years. He died of a heart attack October 9, 1901, at the age of 62.

References Cited:
1. HARTIG, T. 1833. Abhandlung über die Verwandlung der polycotyledonischen Pflanzenzelle in Pilz und Schwammgebilde und der daraus hervorgehenden sogenannten Fäulniss des Holzes. Berlin. 46. p.
2. SCHACHT, H. v. 1863. Ueber die Veränderungen durch Pilze in abgestorbenen Pflanzenzellen. Jahrb. f. wiss. Botanik 3:442-483. (An unedited English translation of this paper is available from the senior author)
3. WILLKOMM, M. 1866. Zur Kenntnis der Roth- und Weiss-fäule. *In*: Die mikroscopischen Feinde des Waldes 1:31-100.

Publications of Robert Hartig

1. 1865. Vergleichende Untersuchungen über den Wachsthumsgang und den Ertrag der Rothbuche und Eiche im Spessart, der Rothbuche im östlichen Wesergebirge, der Kiefer in Pommern und der Weisstanne im Schwarzwalde. Stuttgart, Cotta'sche Buchhandlung, VI u. 75 p.

2. 1868. Die Rentabilität der Fichtennutzholz- und Buchenbrennholzwirthschaft im Harze und im Wesergebirge. Stuttgart, J. G. Cotta'scher Verlag XII u. 199.

3. 1869. Der Pressler'sche Zuwachsbohrer und die Methoden der Zuwachsermittelung. Z. f. Forst- u. Jagdw. 1:110-128.

4. 1869. Das Aussetzen der Jahresringe bei unterdrückten Stämmen. Z. f. Forst- u. Jagdw. 1:471-476.

5. 1869. Mitteilungen über Pilzkrankheiten der Insekten im Jahre 1868. Z. f. Forst- u. Jagdw. 1:476-500.

6. 1870. Zur Lärchenkrankheit. Z. f. Forst- u. Jagdw. 2:356-359.

7. 1870. Das Auftreten der Rhizomorpha in Nadelholzculturen. Z. f. Forst- u. Jagdw. 2:359-361.

8. 1870. Abnorm gebildete Eicheln. Z. f. Forst- u. Jagdw. 2:399-401.

9. 1870. Cetonia aurata in den künstlichen Maikäferbrutstätten. Z. f. Forst- u. Jagdw. 2:401-403.

10. 1870. Bostrichus bidens in Fichten. Z. f. Forst- u. Jagdw. 2:403.

11. 1870. Coleophora lutipenella. Z. f. Forst- u. Jagdw. 2:404-405.

12. 1870. Nepticula (Tinea) Sericopeza Jllg. Z. f. Forst- u. Jagdw. 2:405-406.

13. 1871. Ueber das Dickenwachsthum der Waldbäume. Z. f. Forst- u. Jagdw. 3:66-104.

14. 1871. Ueber den Einfluss verschiedener Raupenvertilgungsmethoden auf die Gesundheit der Kiefer. Z. f. Forst- u. Jagdw. 3:390-396.

15. 1872. Die Begründung der pflanzenphysiologischen Abtheilung der forstlichen Versuchsstation zu Neustadt-Eberswalde. Z. f. Forst- u. Jagdw. 4:96-99.

16. 1872. Caeoma pinitorquum. A. Br. Z. f. Forst- u. Jagdw. 4:99-123.

17. 1872. Benützung des Tamariskenmooses Hypnum tamariscum zur Anfertigung künstlicher Blumen. Z. f. Forst- u. Jagdw. 4:155-159.

18. 1872. Erwiderung auf Herrn Draudts Beurtheilung meines Verfahrens der Holzmassenermittelung. Z. f. Forst- u. Jagdw. 4:160-162.

19. 1872. Einfluss verschieden starker Ausästung und Entnadelung auf den Zuwachs der Weymouthskiefer und gemeinen Kiefer. Z. f. Forst- u. Jagdw. 4:240-254.

20. 1872. Die Misserfolge beim Anbau der kaspischen Weide und das Erkranken derselben durch Melampsora salicina. Z. f. Forst- u. Jagdw. 4:254-263.

21. 1872. Zur Beurtheilung der Lebensfähigkeit der durch Raupenfrass entnadelten Kiefer. Z. f. Forst- u. Jagdw. 4:263-265.

22. 1873. Ueber Rindenproduction der Kiefer. Z. f. Forst- u. Jagdw. 5:195-203.

23. 1874. Das specifische Frisch- und Trockengewicht, der Wassergehalt und das Schwinden des Kiefernholzes. Z. f. Forst- u. Jagdw. 6:194-218.

24. 1874. Wichtige Krankheiten der Waldbäume. Beiträge zur Mykologie und Phytopathologie für Botaniker und Forstmänner. Berlin, J. Springer. 124 p.

25. 1875. Die durch Pilze erzeugten Krankheiten der Waldbäume. 2nd ed., Breslau, Verlag von M. E. Morgenstern. 24 p.

26. 1876. Die Buchencotyledonenkrankheit. Z. f. Forst- u. Jagdw. 8:117-123.

27. 1876. Zur Kenntnis von Loranthus europaeus und Viscum album. Z. f. Forst- u. Jagdw. 8:321-329.

28. 1876. Der Wurzeltödter der Eiche Rhizoctonia quercina. Z. f. Forst- u. Jagdw. 8:329-330.

29. 1876. Ueber die Blitzbeschädigung der Waldbäume. Z. f. Forst- u. Jagdw. 8:330-332.

30. 1877. Die Rotfäule der Fichte. Monatsschr. f. d. Forst- u. Jagdw. 21:97-113.

31. 1878. Die krebsartigen Krankheiten der Rothbuche. Z. f. Forst- u. Jagdw. 9:377-383.

32. 1878. Die Zersetzungserscheinungen des Holzes der Nadelholzbäume und der Eiche in forstlicher, botanischer und chemischer Richtung. Springer, Berlin. VI u. 151 p.

33. 1879. Die anatomischen Unterscheidungsmerkmale der wichtigeren in Deutschland wachsenden Hölzer. 1st ed., 1879, 22 p.; 2nd ed., 1883, 32 p.; 3rd ed., 1890, 40 p.; 4th ed., 1898, 42 p. Rieger, München.

34. 1879. Ein Beitrag zur Eichenästungsfrage. Forstw. Cbl. 1:19-27.

35. 1879. Die Buchenkeimlingskrankheit, erzeugt durch Phytophthora Fagi M. Forstw. Cbl. 1:161-170.

36. 1879. Der Fichtenrindenkrebs, erzeugt durch Nectria Cucurbitula Fr. und Grapholita pactolana Kühlw. Forstw. Cbl. 1:471-476.

37. 1880. Untersuchungen aus dem forstbotanischen Institut zu München. 1. 1880. Springer, Berlin. VIII u. 165 p.

38. 1881. Ueber die durch Pilze bedingten Pflanzenkrankheiten. Ärztl. Intelligenzblatt München 1880. IV u. 29 p.

39. 1881. Ueber Aecidium columnare und Calyptospora Göppertiana Kühn. Flora 39:45.

40. 1881. Anbauversuche mit fremdländischen Holzarten. Flora 39:119-121.

41. 1882. Ueber die Vertheilung der organischen Substanz, des Wassers und Luftraumes in den Bäumen und über die Ursachen der Wasserbewegung in transpirierenden Pflanzen. Berlin, Springer. 112 p.

42. 1882. Lehrbuch der Baumkrankheiten. 1st ed. Springer, Berlin, VIII u. 198 p.; 2nd ed. 1889, Springer, Berlin, IX u. 291 p.; 3rd ed. as "Lehrbuch der Pflanzenkrankheiten" 1900, Springer, Berlin, 324 p.

43. 1882. Ueber den Beginn und den Schluss der Jahrringbildung der Bäume in den verschiedenen Baumhöhen. Flora 40:118-124.

44. 1882. Ueber die normalen Veränderungen des Holzkörpers. Flora 40:543-544.

45. 1882. Bemerkungen zu der statistischen Erhebung über das Vorkommen fremdländischer Waldbäume in Deutschland. Allg. Forst- u. Jagdztg. 58:217-220.

46. 1882. Ueber die wirtschaftliche Bedeutung des sogenannten Vorwuchses bei Begründung und Formbildung reiner und gemischter Waldbestände. Forstwiss. Cbl. 4:1-21.

47. 1883. Untersuchungen aus dem forstbot. Institut zu München. 3 vol. Edited by R. Hartig. Springer, Berlin.
R. Hartig: Zur Lehre von der Wasserbewegung in transpirierenden Pflanzen.
Vervollständigung der Untersuchungen über den Einfluss des Holzalters und der Jahrringbreite auf die Menge der organischen Substanz, auf das Trockengewicht und das Schwinden des Holzes.
Ueber das Verhältnis des lufttrockenen Zustandes der Hölzer zum absolut trockenen Zustand derselben.
Rhizomorpha (Dematophora) necatrix n. sp. Der Wurzelpilz des Weinstockes. Der Wurzelschimmel der Weinreben. Die Wurzelstockfäule.
Das Zerspringen der Hainbuchenrinde nach plötzlicher Zuwachssteigerung.
Erkranken älterer Weymouthskieferbestände.
Mittheilung über Coleosporium Senecionis, den Erzeuger des Kienzopfes.

48. 1883. Die Gasdrucktheorie und die Sachs'sche Imbibitionstheorie. Springer, Berlin. 22 p.

49. 1883. Eine neue Art der Frostbeschädigungen in Fichten- und Tannensaat-und Pflanzenbeeten. Allg. Forst- u. Jagdztg. 59:406-409.

50. 1883. Beschädigung der Nadelholzsaatbeete durch Phytophthora omnivora. Forstwiss. Cbl. 5:593-596.

51. 1883. Eigenthümliche Krankheit der Weymouthskiefer. Flora 41:224-225.

52. 1883. Gewaltsames Zerspringen der Baumrinden nach plötzlich eingetretener Zuwachssteigerung. Flora 41:242.

53. 1883. Ueber die Wasserbewegung in den Pflanzen. Bot. Ztg. 41:250-255.

54. 1883. Die Wasserverdunstung und Wasseraufnahme der Baumzweige im winterlichen Zustande. Flora 41:361-366.

55. 1884. Ein neuer Parasit der Weisstanne, Trichosphaeria parasitica n. sp. Allg. Forst- u. Jagdztg. 60:11-14.

56. 1884. Der Einfluss des Baumalters und der Jahrringbreite auf die Beschaffenheit des Holzes. Allg. Forst- u. Jagdztg. 60:128-137.

57. 1884. Die Ansichten des Herrn Forstrathes Prof. Dr. Nördlinger in Tübingen über die Pilze. Allg. Forst- u. Jagdztg. 60:235-241.

58. 1884. Erscheint es gerechtfertigt, die Forstverwaltungsbeamten zur Vornahme der phänologischen und klimatologischen Beobachtungen zu veranlassen? Allg. Forst- u. Jagdztg. 60:314-316.

59. 1885. Das Holz unserer deutschen Nadelwaldbäume. Springer, Berlin. 147 p.

60. 1885. Die Zerstörung des Bauholzes durch Pilze. 1. Der ächte Hausschwamm (Merulius lacrymans Fr.). Springer, Berlin. 82 p.

61. 1885. Bemerkungen zu einem Vortrage des Herrn Allescher über die südbayerischen Hymenomyceten. Bot. Cbl. 23:362-363.

62. 1885. Die Keimung der Hausschwammsporen. Bot. Cbl. 21:155.

63. 1885. Qualität des Nadelholzes. Bot. Cbl. 23:368-369.

64. 1885. Die Aspe (Populus tremula) als Feind der Kiefer und Lärchenschonungen. Allg. Forst- u. Jagdztg. 61:326-327.

65. 1886. Symbiontische Erscheinungen im Pflanzenleben. Bot. Cbl. 25:350-352.

66. 1886. Die Organisation und Thätigkeit der botanischen Abtheilung der forstlichen Versuchsanstalt in München. Allg. Forst- u. Jagdztg. 62:407-410.

67. 1887. Die Rothstreifigkeit des Bau- und Blochholzes und die Trockenfäule. Allg. Forst- u. Jagdztg. 63:365-368.

68. 1888. Das Holz der Rothbuche in anatomisch-physiologischer, chemischer und forstlicher Richtung. In cooperation with Prof. Dr. R. Weber. Springer, Berlin. VI u. 238 p.

69. 1888. Ueber den Lichtstandzuwachs der Kiefer. Allg. Forst- u. Jagdztg. 64:1-7.

70. 1888. Herpotrichia nigra n. sp. Allg. Forst- u. Jagdztg. 64:15-17.

71. 1888. Der Hausschwamm (Merulius lacrymans). Allg. Forst- u. Jagdztg. 64:49-52.

72. 1888. Die Productionsfähigkeit verschiedener Holzarten auf gleichem Standorte. Allg. Forst- u. Jagdztg. 64:52-55.

73. 1888. Die pflanzlichen Wurzelparasiten. Allg. Forst- u. Jagdztg. 64:118-123.

74. 1888. Die Wasserverdunstung und Wasseraufnahme der Baumzweige im winterlichen Zustande. Flora 41:361-366.

75. 1888. Ueber den Einfluss der Samenproduktion auf Zuwachsgrösse und Reservestoffvorrat der Bäume. Bot. Cbl. 36:388-391.

76. 1888. Ueber die Bedeutung der Reservestoffe für den Baum. Bot. Z. 46:837-842.

77. 1888. Das Fichten- und Tannenholz des bayerischen Waldes. Cbl. f. d. gesamte Forstwesen 14:357-364, 437-442.

78. 1888. Ueber die Wasserleitung im Splintholze der Bäume. Ber. d. deutsch. bot. Ges. 6:222-225.

79. 1888. Die pflanzlichen Wurzelparasiten. Cbl. f. Bakt. 3:118-120.

80. 1888. Zur Verbreitung des Lärchenkrebspilzes Peziza Willkommii. Hedwigia 27:55-58.

81. 1889. Ueber den Einfluss der Samenproduction auf Zuwachsgrösse und Reservestoffvorrath der Bäume. Allg. Forst- u. Jagdztg. 65:13-17.

82. 1889. Die Oberberghauser Weidenanlagen bei Freising. Allg. Forst- u. Jagdztg. 65:48-52.

83. 1889. Die krebsartigen Erkrankungen der Pflanzen. Allg. Forst- u. Jagdztg. 65:120-124.

84. 1889. Ein Ringelungsversuch. Allg. Forst- u. Jagdztg. 65:365, 401.

85. 1889. Zur Kenntniss des Wurzelschwammes Trametes radiciperda. Z. f. Forst- u. Jagdw. 21:428-432.

86. 1889. Die Fällungszeit der Nadelholzbäume im Gebirge. Z. f. Forst- u. Jagdw. 21:457-462.

87. 1889. Bemerkungen zu A. Wieler's Abhandlung: Ueber den Ort der Wasserleitung im Holzkörper. Ber. d. deutsch. bot. Ges. 7:89-94.

88. 1889. Zweijährige Buchenausschläge mit Bucheckern. Bot. Cbl. 37:79.

89. 1889. Eine Krankheit der Weisstanne. Bot. Cbl. 37:78.

90. 1890. Das Studium der Botanik an forstlichen Lehranstalten. Z. f. Forst- u. Jagdw. 22:193-196.

91. 1890. Eine Krankheit der Fichtentriebe. Z. f. Forst- u. Jagdw. 22:667-670.

92. 1890. Was ist in den europäischen Staaten von Seiten derselben bis jetzt gethan worden, um die Erforschung der in forstlicher Hinsicht wichtigen Pflanzenkrankheiten zu fördern und die zerstörenden Wirkungen derselben zu reduciren, und was muss in solcher Richtung noch gethan werden? Cbl. f. d. ges. Forstw. auch Öster. Vierteljahrschr. f. d. Forstw. 4, Vienna.

93. 1891. Lehrbuch der Anatomie und Physiologie der Pflanzen unter besonderer Berücksichtigung der Forstgewächse. Springer, Berlin. VIII u. 308 p.

94. Eine Krankheitserscheinung der Fichtentriebe durch Septoria parasitica. Bot. Cbl. 45:137-138.

95. 1891. Untersuchungen über Rhizina undulata. Bot. Cbl. 45:237-238.

96. 1891. Rostform Melampsora. Bot. Cbl. 46:18.

97. 1891. Ueber die Kelbahn'sche Abhandlung über die Formen des Peridermium Pini. Bot. Cbl. 46:18-19.

98. 1891. Ueber das Verhalten der Fichte gegen Kahlfrass durch die Nonnenraupe. In: A. Pauly, Die Nonne (Liparis monacha) i.d. bayerischen Waldungen. 1890 Frankfurt a.M., 1891 appendix.

99. 1892. Das Erkranken und Absterben der Fichte nach der Entnadelung durch die Nonne (Liparis monacha). Forstl.-naturwiss. Z. 1:1-13, 49-62.

100. 1892. Vertrocknen und Erfrieren der Kieferzweige. Forstl.-naturwiss. Z. 1:85-88.

101. 1892. Niedere Organismen in Raupenblute. Forstl.-naturwiss. Z. 1:124-125.

102. 1892. Ueber den Wuchs der Fichtenbestände des Forstenrieder und Ebersberger Parkes bei München. Forstl.-naturwiss. Z. 1:129-140.

103. 1892. Ueber den Entwicklungsgang der Fichte im geschlossenen Bestande nach Höhe, Form und Inhalt. Forstl.-naturwiss. Z. 1:169-176.

104. 1892. Die Verschiedenheiten in der Qualität und im anatomischen Baue des Fichtenholzes. Forstl.-naturwiss. Z. 1:209-233.

105. 1892. Einfluss der Leimringe auf die Gesundheit der Bäume. Forstl.-naturwiss. Z. 1:281-284.

106. 1892. Ueber das Verhalten der von der Nonne nicht völlig entnadelten Fichten. Forstl.-naturwiss. Z. 1:284-285.

107. 1892. Septogloeum Hartigianum Sacc. n. sp. Ein neuer Parasit des Feldahorns. Forstl.-naturwiss. Z. 1:289-291.

108. 1892. Rhizina undulata Fr. Der Wurzelschwamm. Forstl.-naturwiss. Z. 1:291-297, 477.

109. 1892. Die Erhitzung der Bäume nach völliger oder theilweiser Entnadelung durch die Nonne. Forstl.-naturwiss. Z. 1:369-375.

110. 1892. Ueber die bisherigen Ergebnisse der Anbauversuche mit ausländischen Holzarten in den bayerischen Staatswaldungen. Forstl.-naturwiss. Z. 1:401-414, 415-432, 441-452.

111. 1892. Ein neuer Keimlingspilz. Forstl.-naturwiss. Z. 1:432-436.

112. 1892. Weitere Mittheilungen über die Temperatur der Bäume. Forstl.-naturwiss. Z. 1:475-477.

113. 1892. Ueber Dickenwachsthum und Jahresringbildung. Bot. Ztg. 50:176-180, 193-196.

114. 1893. Eine krebsartige Rindenkrankheit der Eiche, erzeugt durch Aglaospora Taleola. Forstl.-naturwiss. Z. 2:1-6.

115. 1893. Cecidomyia Piceae n. sp., Die Fichtengallmücke. Forstl.-naturwiss. Z. 2:6-8, 274-275.

116. 1893. Der Wachsthumsgang der Fichte im bayerischen Walde. Forstl.-naturwiss. Z. 2:49-57.

117. 1893. Die Spaltung der Oelbäume. Forstl.-naturwiss. Z. 2:57-63.

118. 1893. Wachsthumsgang und Holz der canadischen Pappel. Forstl.-naturwiss. Z. 2:89-93.

119. 1893. Wachsthumsgang und Holz der Robinie. Forstl.-naturwiss. Z. 2:93-96.

120. 1893. Betrachtungen über das forstliche Unterrichts- und Versuchswesen. Forstl.-naturwiss. Z. 2:158-174.

121. 1893. Ein Waldspiel. Forstl.-naturwiss. Z. 2:184-187.

122. 1893. Beschädigung der Bäume durch Leimringe. Forstl.-naturwiss. Z. 2:187.

123. 1893. Untersuchungen über Wachsthumsgang und Ertrag der Eichenbestände des Spessarts. Forstl.-naturwiss. Z. 2:249-269, 289-305.

124. 1893. Ueberblick über die Folgen des Nonnenfrasses für die Gesundheit der Fichte. Forstl.-naturwiss. Z. 2:345-357.

125. 1893. Septoria parasitica M. in älteren Fichtenbeständen. Forstl.-naturwiss. Z. 2:357-359.

126. 1893. Ueber das Verhalten der ausländischen Holzarten zur Kälte des Winters 1892/93. Forstl.-naturwiss. Z. 2:411-413, 460-463.

127. 1894. Untersuchung über die Entstehung und die Eigenschaften des Eichenholzes. Forstl.-naturwiss. Z. 3:1-13, 49-68, 172-191, 193-203.

128. 1894. Sonnenrisse und Frostrisse an der Eiche. Forstl.-naturwiss. Z. 3:255-260.

129. 1894. Die Ausschlagfähigkeit der Eichenstöcke und deren Infection durch Agaricus melleus. Forstl.-naturwiss. Z. 3:428-433.

130. 1894. Untersuchungen des Washsthumsganges der Eiche im Guttenberger und Gramschatzer Walde bei Würzburg und im Forstamt Freising und Starnberg bei München. Forstl.-naturwiss. Z. 3:482-522.

131. 1895. Doppelringe als Folge von Spätfrost. Forstl.-naturwiss. Z. 4:1-8.

132. 1895. Untersuchungen des Baues und der technischen Eigenschaften des Eichenholzes. Forstl.-naturwiss. Z. 4:49-82.

133. 1895. Ueber den Drehwuchs der Kiefer. Forstl.-naturwiss. Z. 4:313-326.

134. 1895. Ueber die Güte des Nonnenholzes. Forstl.-naturwiss. Z. 4:369-374.

135. 1895. Das Absterben der Kiefer nach Spannerfrass. Forstl.-naturwiss. Z. 4:396-404.

136. 1895. Der Nadelschüttepilz der Lärche (Sphaerella laricina n. sp.). Forstl.-naturwiss. Z. 4:445-457.

137. 1896. Wachsthumsuntersuchungen an Fichten. Forstl.-naturwiss. Z. 5:1-15, 33-45.

138. 1896. Ueber das Verhalten der vom Spanner entnadelten Kiefern im Sommer des Jahres 1895. Forstl.-naturwiss. Z. 5:59-64.

139. 1896. Ueber die Einwirkung schwefliger Säure auf die Gesundheit der Fichte. Forstl.-naturwiss. Z. 5:65-69.

140. 1896. Sphaerella laricina auf Larix leptolepis. Forstl.-naturwiss. Z. 5:74.

141. 1896. Das Rothholz der Fichte. Forstl.-naturwiss. Z. 5:96-109, 157-169.

142. 1896. Ueber die Einwirkung des Steinkohlenrauches auf die Gesundheit der Nadelholzbäume. Forstl.-naturwiss. Z. 5:245-290.

143. 1896. Die Folgen des 1895er Spannerfrasses im Nürnberger Reichswalde. Forstl.-naturwiss. Z. 5:311-313.

144. 1896. Die Tannennadelmotte Argyresthia fundella F.R. Forstl.-naturwiss. Z. 5:313-316.

145. 1896. Innere Frostspalten. Forstl.-naturwiss. Z. 5:483-488.

146. 1897. Waldbeschädigung durch ein Eisenwerk. Forstl.-naturwiss. Z. 6:40-44.

147. 1897. Ueber den Einfluss des Hütten- und Steinkohlenrauches auf den Zuwachs der Nadelholzbäume. Forstl.-naturwiss. Z. 6:49-60.

148. 1897. Untersuchungen über Blitzschläge in Waldbäumen. Forstl.-naturwiss. Z. 6:97-120, 145-165, 193-206.

149. 1897. Tödtung der Bucheckern im Winterlager durch Mucor mucedo. Forstl.-naturwiss. Z. 6:337-339.

150. 1897. Verkohlung der Lärchenborke im Hochgebirge. Forstl.-naturwiss. Z. 6:473-474.

151. 1897. Ueber den Einfluss der Erziehung auf die Beschaffenheit des Holzes der Waldbäume. Schweiz. Z. f. Forstw. 48:93-98.

152. 1898. Bau und Gewicht des Fichtenholzes auf bestem Standorte. Forstl.-naturwiss. Z. 7:1-19.

153. 1898. Ueber den Einfluss der Kronengrösse und der Nährstoffzufuhr aus dem Boden auf Grösse und Form des Zuwachses und auf den anatomischen Bau des Holzes. Forstl.-naturwiss. Z. 7:73-94.

154. 1898. Die Grafrather Anbauversuche. Forstw. Cbl. 20:373-378.

155. 1899. Ueber die Ursachen exzentrischen Wuchses der Waldbäume. Cbl. f. d. ges. Forstw. 25:292-307.

156. 1899. Neue Beobachtungen über Blitzbeschädigungen der Bäume. Cbl. f. d. ges. Forstw. 25:360-381, 523-544.

157. 1899. Die Lärchennadelbräune, erzeugt durch Allescheria Laricis n. sp. Cbl. f. d. ges. Forstw. 25:423-426.

158. 1899. Phoma sordida Sacc., ein neuer Hainbuchenparasit. Cbl. f. d. ges. Forstw. 25:485-486.

159. 1900. Beiträge zur Kenntnis des Eichenwurzeltödters (Rosellinia quercina M.) Cbl. f. d. ges. Forstw. 26:243-250.

160. 1901. Holzuntersuchungen. Altes und Neues. Springer, Berlin. VI u. 99 p.

161. 1901. Ueber die Borkenbildung des Bergahorns. Cbl. f. d. ges. Forstw. 27:49-51.

162. 1901. Ueber die Ursachen des Wimmerwuchses. Cbl. f. d. ges. Forstw. 27:145-150.

163. 1901. Agaricus melleus, ein echter Parasit des Ahorns. Cbl. f. d. ges. Forstw. 27:193-196.

164. 1902. Der echte Hausschwamm und andere das Bauholz zerstörende Pilze. 2nd ed. edited by von Tubeuf, Springer, Berlin. 105 p.

Wichtige
Krankheiten der Waldbäume.

Beiträge zur Mycologie und Phytopathologie

für

Botaniker und Forstmänner

von

Dr. Robert Hartig,

Professor der Botanik an der Königl. Preuss. Forstakademie zu Neustadt-Eberswalde und Vorstand der pflanzenphysiologischen
Abtheilung des forstlichen Versuchswesens in Preussen.

Mit 160 Originalzeichnungen auf 6 lithographirten Doppeltafeln.

BERLIN, 1874.

Verlag von Julius Springer.
Monbijouplatz 3.

IMPORTANT

DISEASES OF FOREST TREES

Contributions to Mycology and Phytopathology

for

Botanists and Foresters

by

Dr. Robert Hartig

Professor of Botany at the Royal Prussian Forestry Academy at Neustadt-Eberswalde
and Director of the Plant Physiology Section of the Forest Research Organization
in Prussia

With 160 Original Illustrations on 6 Lithographed Double Plates

Berlin, 1874

Publishing House of Julius Springer

Monbijouplatz 3

English translation by
William Merrill, David H. Lambert, and Walter Liese

FOREWORD

I hereby submit to the botanical and forestry public a series of investigations which have resulted from endeavors to lift a little the veil which until now has lay spread over most diseases of forest trees. Although my practical and scientific* career as a forester has reached its end because of the appointment to the Botany Chair at the Forestry Academy of Neustadt-Eberswalde, I believe the following work to give proof that I have remained a forester and have become a botanist.

I am well aware of the danger of such a dual position, but believe that the success which I have reached after a short transition period must be due above all to this position.

Whereas for a long time the diseases of agriculturally cultivated plants have been investigated and explained, of the most wide spread diseases of forest trees only a few have been investigated at all and even these few in no way elucidated in a satisfying manner.

Botanists lack the opportunity to observe and to experiment in the forest; on the other hand, forestry people lack the necessary botanical knowledge since, in fairness, a basic study of mycology is not required of a forester.

It was obvious that a great number of diseases confronted me immediately, most of which had not at all and others had only insufficiently been described and investigated.

The first series of my pathological works I publish now in the hope that it will be possible for me to allow others to follow in a short time.

If with only a number of diseases, to which certainly just the most important belong, I can arrive at important results for the practical forester, if I can suggest means which can be brought into use against these, then the explanation of the causes and appearances of these diseases will give satisfaction to the educated forester even if practical results cannot immediately be drawn therefrom.

By the same right by which we expect that the forester knows the harmful and useful insects of his forest and their life cycles, one can also demand that he be able to form an opinion about the diseases of his trees and their causes even if he has no means at his disposal to take steps against them.

For the benefit of the foresters, I have given in the introduction a summary of the morphology and physiology of fungi with reference to the explanatory illustrations.

In order not to hinder the understanding of the following investigations by too much material, all which did not appear necessary is deleted.

---

*R. Hartig; Die Rentabilität der Fichtennutzholz- und Buchenbrennholzwirthschaft. Stuttgart, Cotta. 1868.

He who is deeply interested in the study of fungi, I refer to the masterful presentation in: de Bary: Morphology and Physiology of the Fungi, Lichens, and Slime Molds, Leipzig, Engelmann, 1866, which I have taken for a basis of this review.

For the mycologist the introduction has interest only insofar as I have drawn attention in this to some of my observations which are of general mycological interest.

With the great abundance of new observations, it is important to facilitate as much as possible the opportunity for controlled investigations and therefore I fulfilled with pleasure the wish of Herr Baron von Thümen to contribute the described parasites for his "Herbarium mycologicum oeconomicum," as much as it was possible for me to supply these immediately in such ample amounts that they are sufficient for an entire edition of this essicatti which appeared only recently in the book trade.

Most of the parasites described by me occur in Fascicle 3 under the following numbers: *Rhizomorpha subterranea* No. 143; *Rhizomorpha subcorticalis* No. 192; *Trametes Pini* fruit body No. 136; *Peridermium Pini corticola* No. 190; *Peridermium Pini acicola* No. 140; *Caeoma pinitorquum* No. 139; *Caeoma Laricis* No. 189; *Peziza Willkommii* No. 191; *Hypoderma macrosporum* No. 79, Fascicle 2 (this erroneously named *Hypoderma longisporium*); *Hypoderma nervisequium* No. 43, Fascicle 1; *Melampsora salicina* No. 142.

Those still lacking perhaps will be added in later fascicles. For mycologists who wish to have more abundant material to investigate, I am ready to send these with pleasure and I will check most conscientiously and, if necessary, thankfully accept each correction or completion of my reports.

Scientifically nonfounded differences of opinion which so often spring from the prejudices of envious distinguished forester colleagues, if they deal with questions concerning fungi, I will not answer.

Neustadt-Eberswalde in October, 1873.

Robert Hartig

CONTENTS

INTRODUCTION

Review of the Morphology and Physiology of Fungi

1. Mycelium

In each fungus plant one distinguishes two parts, which are more or less decisively differentiated by form and function. The nutrition absorbing and distributing part, the so-called mycelium, is as a rule concealed within the organism or substance from which the fungus absorbs its nutrients, and therefore usually is not at all known to those unacquainted with fungi. Only that mycelium which develops on the surface of the nutrient substrate is recognized and bears the collective and common name "mold" (Schimmel). The second part of the fungus plant, the fruit body with the reproductive organs produced therein, is better known and is more conspicuous by size and form. Only the mysterious group of Schizomycetes, to which belong the single celled bacteria, vibrios, and so forth, do not permit a differentiation between mycelium and fruit body, since they consist only of simple cells which increase by division.

Since the fungi, like all other organisms, never arise through spontaneous generation but only from germination of previously formed organs of the same species, so can we only observe these in the most youthful condition when we germinate reproductive cells. We see germinating spores (Plate II, Fig. 26; Plate III, Fig. 12; Plate IV, Fig. 18; Plate V, Fig. 9; Plate VI, Fig. 12, 16, 24, 27, 28).

The young fungus plant which arises from the reproductive cell or spore consists of a slender, usually colorless tube, which enlarges by means of growth of the tip and by ramifying branches which bud out laterally behind the tip. The content of the tube is originally very rich in plasma and usually shows larger or smaller oil drops, which often possess a gold-yellow color. Later, as a rule, the clear cell sap exceeds the plasma, which is crowded against the cell wall, or the fluid cell contents disappear completely and are replaced by air. In most young fungus tubes (known as hyphae) inner crosswalls arise by which the mycelium is divided into a series of cylindrical cells. (Plate I, Fig. 12; Plate III, Fig. 13, etc.) The young germinated plant, which has lived only a brief time at the expense of the nutrient materials which are contained in the spore, when it encounters conditions which are favorable for its nourishment, develops by a lengthening of the hypha from the tip and by sending out lateral hyphae and thereby forming the mycelium which, by its nutritional processes, brings about the most important changes in organic bodies, which we will learn about later. The following special investigations together with new observations furnish an abundant review of the diversity of form of mycelium. In the predominating majority of cases the hyphae or mycelial threads remain isolated and do not come together into a larger fungus body, forming a so-called "simple threaded mycelium". The cell wall of the fungus hypha, in the beginning always very thin, often later becomes greatly thickened so that the lumen becomes reduced to a fine canal. Most interesting in this respect is the mycelium of *Hypoderma macrosporum* (Plate VI, Fig. 6). In a youthful condition (a) showing a fine, nevertheless already clearly double layered wall, later the mycelium shows three layers (Fig. 6b, Fig. 7b), of which the very thick and finely layered middle one is colored intensely blue by iodine. The hyphae which are gathered together into a pseudoparenchyma of the cord-shaped mycelial body of *Agaricus melleus,* the so-called rhizomorph, are in their outer part, in the so-called rind of the strand, so greatly thickened that the lumen almost completely vanishes. The outer boundary is clearly detectable only after treatment with KOH (Plate I, Fig. 18, 19). Often hyphae and spores are surrounded by a transparent jelly (Plate VI, Fig. 9, 11, 12, 13, 24), or numerous fungus threads lie embedded in a common jelly which can be regarded as a secretion product of them (Plate I, Fig. 10, 11, 12, 26c). The hairs in the rind of *Rhizomorpha* (Plate I, Fig. 8; Plate II, Fig. 9) show an extremely fine grained secretion on the surface of the cell wall.

The hyphae on the surface of *Rhizomorpha subcorticalis* which are embedded in jelly perhaps one centimeter behind the growing point often have a highly characteristic wall thickening which can be explained only by an elongation of the hypha at a time when the cell wall no longer is pliable. The fragments of the latter are held together by the plasmalemma or Ptychodeschlauch (Plate I, Fig. 13).

The mycelial threads which grow in plant tissue are either intercellular, that is, they grow between the cell walls of the parenchyma cells or the prosenchyma, grow abundantly in the resin canals, in the intercellular spaces of the leaf spongy parenchyma, sending at most into the interior of the cells short branches, the so-called haustoria (Plate IV, Fig. 19-22; Plate V, Fig. 6, 8, 12, etc.; Plate VI, Fig. 5) or they grow in the interior of the cells, boring through the cell walls, partly to pass from one cell to another, partly to withdraw nutrient materials from the wall and to carry out a more rapid decomposition (Plate I, Fig. 14; Plate III, Fig. 16, 17, 18).

The "simple threaded mycelium" under favorable conditions, when it encounters cavities for freer development in the interior of the tree, fills these cavities with a more or less thick, usually leatherlike skin (Plate III, Fig. 1g, 17) or nonuniform mycelial bodies (Plate III, Fig. 1e) which show no characteristics but consist only of thick intertwining mycelial threads (Plate III, Fig. 17a). In earlier times these were described as specific fungi of the genus *Xylostroma.*

Of unusually peculiar structure are some mycelial bodies which have either a knobby form and are known as sclerotia (Mutterkorn), or produce root-like strands and broad, flattened bands.

Of these so-called rhizomorphs, I have subjected the most frequently occurring species, *Rhizomorpha fragilis* Roth, to a very thorough investigation, the results of which are shown in Plate I, while the fruit bodies which belong to this mycelium are illustrated in Plate II. The first special discussion gives details concerning the structure of this mycelial body.

2. The Fruit Bodies and Reproductive Organs

The second part of the fungus plant, which has the function of producing and bearing the reproductive organs, is known as the fruit body (Receptaculum). Only the larger forms of the fruit bodies are generally known and are called mushrooms (Schwämme), while the very interesting microscopic fruit bodies with diversity of form tend to remain unobserved.

Sometimes the fruit bodies consist only of simple branches of the mycelium, on whose ends the reproductive cells (spores) are constricted off or are produced in special sporocarps (Sporangien) (Plate VI, Fig. 17). These fruiting hyphae are known as conidiophores. Usually numerous mycelial threads join together here and there for the formation of composite sporophores, which then consist of a thick texture of numerous interlaced fungus threads (Plate II, Fig. 10m; Plate III, Fig. 8, 9, 10). If the hyphae which form the fruit body run more or less parallel and are crowded closely together, then a section made perpendicular to the longitudinal direction of the tissue often furnishes an image which possesses the greatest similarity to a parenchymatous tissue and is known as pseudoparenchyma (Plate I, Fig. 19; Plate V, Fig. 13p, 15st, etc.).

The form and development of composite fruit bodies is extremely varied. Plate II shows the development of large stemmed and veiled fruit bodies of *Agaricus melleus* from their first beginnings.

Likewise, Plate III shows the development of the unstalked bracket formed conk (Schwämme) of *Trametes*. The cup-shaped fruit bodies of *Peziza* and *Corticium* (Plate V, Fig. 16, 18, 20), which expand flatly, and the fruiting bodies of the Uredinales and Discomycetes, which are buried in substance of the host plant (Plate IV, Fig. 19; Plate V, Fig. 5, 6, 14; Plate VI, Fig. 10, 13, 21, etc.), are represented frequently among the parasites of forest trees.

The cells which produce spores in the fungi, which are of interest to us here, form a complete layer in the sporophore, which is known as the spore layer, hymenium or fruiting layer. Usually this covers only a certain part of the sporophore; for example, with *Agaricus* the outer surface of the lamella which arises on the under side of the cap (Plate II, Fig. 22, 29), in the case of *Trametes* on the walls of the pores (Plate III, Fig. 8, 9, 10, 24, 25), in the case of *Peziza* and *Corticium* in the inner cavity of the cups (Plate V, Fig. 16, 18, 20).

In the case of the Uredinales and Discomycetes (Plates IV, V, VI), the entire outer side of the sporophore is covered with the hymenial layer.

In these the reproductive cells (spores) arise in different manners. In the case of many fungi, the Ascomycetes, the hymenial layer consists of freely standing club-shaped tubes (asci) surrounded by numerous hair-shaped hyphae, paraphyses, which develop before the formation of asci in the receptaculum. In these tubes usually eight spores arise by free cell formation, which germinate after their liberation (Plate V, Fig. 18-21; Plate VI, Fig. 9, 10, 11, 23).

Free cell formation in the asci is in contrast with the formation of spores by abstriction from the top of the spore mother cell (basidium).

The process of abstriction takes place only once on the mother cell (simultaneous) or a large number of spores arise one after another by repeated abstriction from the same basidium (successive).

We see simultaneous abstriction in the Hymenomycetes, for example, in *Agaricus melleus* (Plate II, Fig. 22, 23), in *Trametes* (Plate III, Fig. 9, 10, 25) and *Corticium* (Plate V, Fig. 17).

In the case of the latter, the basidia are surrounded by paraphyses which are constricted beadlike on the end; it seems doubtful to me whether the numerous nonsporulating club-shaped cells in the hymenial layer of *Agaricus* (Plate II, Fig. 22) should be considered to be paraphyses. I consider them to be sterile basidia and those thin lineal cells which still occur beside them to be rudimentary basidia (Hoffman's Pollinarien) (Plate II, Fig. 22, 23c) especially since under favorable conditions they can also germinate just like spores (Plate II, Fig. 24).

In the case of the previously mentioned fungi, there arise on the tips of the basidia usually four awl-shaped outgrowths, sterigma, which swell up bubblelike at the end until the spores thus arising have attained their normal size.

Basidia which cut off spores in succession are illustrated in Plate IV, Fig. 14, 16, 19 and Plate V, Fig. 8 and 14. The spores which are formed from a basidium are at first upright in a row, frequently are separated from one another by so called disjunctor cells (Zwischenstücke), which, with the complete development of the spore wall, are reabsorbed and disappear, therefore perhaps serve for the development of the spores.

When the spores are fully developed they usually fall from the basidium and are disseminated by the wind.

Processes of copulation and sexual reproduction as have been demonstrated with many fungi were not observed in the following investigations; however, the spermogonia, which are considered to be the male sexual organs, were found commonly with the spermatia which are produced thereon. They are described for *Peridermium Pini* (Plate IV, Fig. 1, 2, 7, 15, 16, 20), for *Caeoma pinitorquum* (Plate V, Fig. 4, 5, 6), *Caeoma Laricis* (Plate V, Fig. 11, 12, 13, 15), for *Hypoderma macrosporum* (Plate VI, Fig. 13, 15), *Hypoderma nervisequium* (Plate VI, Fig. 20, 21, 22). The illustrated form of the spermatia of *Hypoderma macrosporum* (Plate VI, Fig. 14, 15) germinate (Plate VI, Fig. 16), and develop a filamentous fungus (Plate VI, Fig. 17) which cuts off conidia.

It is, therefore, quite possible that those organs should not be considered to be spermatia but to be conidia (spores).

Whether the bubble-shaped swellings of the hyphae which I have observed on some rhizomorph ends (Plate I, Fig. 12) as well as similar formations on the paraphyses of *Hypoderma nervisequium* (Plate VI, Fig. 25) bear any relationship to the fructification process I must leave undetermined.

Size and shape of spores is exceedingly varied. The plates display round to filiform shapes with various forms in between. The wall is extremely thin or very thick often with a differentiation between an outer and inner layer, the epispore or exospore, and the endospore.

The exospore often shows characteristic wart-shaped thickenings or is surrounded by a gelatinous coat. In very thick walls one can often detect one or more germ pores (Plate VI, Fig. 27, 28) from which the germ tube emerges.

Most spores germinate with only one germ tube, others in very favorable conditions produce two or three tubes at the same time (Plate VI, Fig. 27), etc.

It is noteworthy that with a variation of external conditions the germ tube is able to take on a different form. The urediospores of *Melampsora salicina* artificially brought to germination in moist air on a microscope slide produce thick tubes (Plate VI, Fig. 28), while those germinating in free air on the upper surface of willow leaves send forth very fine germ tubes (Plate VI, Fig. 27).

3. Polymorphism

From the germinating spore a new fungus plant arises which in many cases is completely dissimilar to the mother plant and which produces sporophores and spores completely different from the first form.

Several fungi even produce a whole series of forms which, in the developmental cycle of the plant, can or must occur one after the other, before they again produce the original form. Polymorphism of the fungi could be compared quite appropriately with that of insects if, however, the caterpillar, after it had ended its development, layed eggs (spores) from which small pupae developed, which later grew larger, if the pupae likewise layed eggs which produced small butterflies which developed further. Just as in the case of the insects where that developmental stage in which the sexual developmental process occurs is labeled as the highest, so also in the fungi one denotes that form of reproduction in which sexual processes are detected or are assumed as probable as the highest development and calls it "fructification." As such most highly developed forms one considers the sporophores which bear asci, the aecia and so forth. The remaining forms of a fungus species, on the contrary, one denotes as "propagation." In a number of fungi the organs of propagation (asexual reproduction) are not necessary members in the development of the plant, in certain cases they can be absent. The conidia of *Hypoderma nervisequium* (Plate VI, Fig. 21, 22) perhaps belong here, since these develop regularly only under special climatic conditions. In most fungi, on the contrary, the propagation cells are necessary developmental components and are never absent, indeed, they often are absolutely essential to facilitate the alternation of generations of the fungus. In addition to the urediospore form,

Lecythaea salicis, which develops in the summer on the leaves and in the bark of *Salix acutifolia* (Plate VI, Fig. 26), belongs *Melampsora salicina*, which overwinters on the fallen blackening leaves, whose resting spores germinate in the spring, and produce a small promycelium, on which arise numerous very small spores (sporidia). If these reach the healthy leaves of willow, they can again produce the disease. The alternation of generations between *Aecidium Berberidis* on barberry leaves and the cereal rust fungus, *Uredo linearis,* as well as the overwintering resting spores of *Puccinia graminis* need here to be only recalled, since these are well enough known to the reader.

Often it is a case of a fortunate accident to discover in which different forms and on which different host plants a fungus occurs, although we will never have perspective of the results of our investigations if we do not, like officers of the court, carefully consider all suspicious features which are able to lead us to the right trail. All plants in the vicinity of an infected stand and the fungi which are growing upon them are to be investigated carefully and placed under continued inspection. The occurrence of disease under certain external conditions in the vicinity of the field, on certain soil types, in certain exposures, and so forth perhaps is able to give us a hint for further final investigations, which result, in our success, by infection of the host plant with the suspected fungus, in producing the second associated form of the fungus and in producing the disease. With certainty we can assume that still other fungus forms belong to *Peridermium Pini, Caeoma pinitorquum* and *C. Laricis* whose confirmation will be the task of further investigations.

4. Physiology of Fungi

Fungus spores germinate only under certain conditions which are very different for individual species and lack of knowledge of these so often explains the failure of germination and infection studies.

Just as with the seeds of phanerogamic plants the period of seed dormancy is very different, thus also spores of fungi often germinate only after a long spore dormancy, often immediately after maturity. Germinability sometimes lasts several years, sometimes only a very short time.

The spores of *Agaricus melleus*, taken from a pileus which had just opened, germinated with great ease; those spores taken from a pileus which had been opened for several days appeared to have completely lost the ability to germinate.

The minimal germination temperature for most fungi apparently lies well below that of the higher plants; how high a temperature the spores can bear has been investigated for only a few species.

In water perhaps all spores are killed at 100° C while it is proven that many spores can withstand a temperature of 110° to 120° C in dry air without being killed.

Suitable temperature and moist air or liquid water often bring spores to germination while in other cases special conditions, above all a certain nutrition, are necessary. Moist air or damp atmospheric conditions hasten the

[6]

germination of spores just like the development of fungi even if they are still buried within the interior of the host plant.

This explains many phenomena about the diseases of plants which in earlier times were regarded as direct results of climatic influences or of the soil.

In damp areas, in areas less exposed to ventilation, in wet years, in the lower parts of tree crowns and so forth, diseases produced by fungus parasites develop more severely than where the germination of spores and development of fungi is impaired by drier air. Mycelium of *Caeoma pinitorquum* becomes perennial in *Pinus sylvestris* once it is attacked by the fungus and moves annually into the new shoots to fructify in their bark. With wet weather the disease appears devastatingly; the abundant development of the fungus kills a great number of the shoots. In dry and hot years one sees on the shoots only occasional yellowish spots on the bark; the fruiting bodies remain in the first developmental stage or become stunted, without having brought the plant the least harm.

Since fungi in general do not possess the ability to assimilate inorganic materials (some exceptions in the case of the yeast fungi will remain unconsidered here) they lack the need for chlorophyll and light. They grow in the deep darkness of tree interiors. Like animals, they need large amounts of oxygen which they withdraw from the air or, when this is cut off, from the nutrient materials, secreting carbonic acid for it. Nitrogen-rich nutrition benefits their growth since the plasma contents of the mycelial strands are very rich in proteinaceous materials. If the nutritive body is very poor in proteinaceous substances, as for example wood, then the fungi manage such that the older parts of the mycelial threads very soon lose their plasma to the growing tip, the plasma of the hyphal tip follows in a similar fashion, the parts lying behind empty themselves and soon die, indeed, often are resorbed early and disappear. In ring-shake pine wood numerous bore holes in the fiber walls indicate although many fungus hyphae have penetrated into the wood fibers in the course of the disease, often no more trace of them is to be found (Plate III, Fig. 18).

According to nutrition the fungi were at earlier times divided into two groups, into rot inhabitors (saprophytes) which live on already killed animals and plants and bring about and hasten their destruction, and into parasites (Schmarotzer), which attack living organisms, obtain their nutrition from these, whereby these become diseased or are killed. Such a strict separation is not always feasible, since most parasites also feed from the substances of the plants or animals after these have been killed by them.

Thus, for example, the mycelium of *Hypoderma macrosporum* (Plate VI, Fig. 5) kills the leaf parenchyma of spruce needles (at m). As a result of this, these shrivel up and lose their green color (n) while the starch granules are still unchanged. These fungus hyphae also cause the further destruction and decomposition of the killed parenchyma cells, which at (o) contain no more starch.

Most parasites in the second stage of their development are saprophytes. An especially interesting case is the life cycle of the mycelium of *Agaricus melleus*. *Rhizomorpha fragilis* is a fungus mycelium which is widely distributed on killed hardwood roots and stems, on construction wood in bridges, water pipes, and mines, and lives here only as a saprophyte.

This fungus is a true parasite only on the firs which bear resin canals and apparently also on *Prunus* species; it is able to kill within a short time completely healthy trees in their most vigorous growth.

The significance of saprophytes in the household of nature was known quite early, even although not in its entirety. It is known that each process which is distinguished with the expression: fermentation, rot, molding, decay, humification, and so forth, is destruction of organic substances brought about or at least initiated by the life processes of fungi or certain lower animals. The rapid and definite occurrence of the processes of destruction is explained by the fact that innumerable microscopically small spores or mycelial pieces float around in the air, which, falling on a fermentable body, soon develop. It has been claimed that one searches in vain for such fungi in forest air. I therefore placed an aspirator in the forest, in which the forest air must be sucked through between two closely connected small glass plates. The space between the two glass plates was filled with carefully sterilized cotton, in the threads of which the spores, which occurred in the air flowing through, remained hanging.

After a filtration of only ten liters of air, numerous fungus spores could be detected under the microscope; these appear still more clearly if one allows some distilled water to be drawn between the glass plates. If one allows these to lie only a few days after moistening, one sees an abundant fungus vegetation issuing forth from the moistened and germinated spores.

Fungi withdraw from organic substances those materials which they need for their existence.

In a great extent this pertains to oxygen, which is withdrawn from the nutrient material, if the air does not have full access to the fungus. The rest of the organic substance which remains behind cannot remain unchanged, but decomposes into simpler chemical compounds; this process is called fermentation. The end products of fermentation are water, carbonic acid, different hydrogen compounds, such as methane or hydrogen sulfide, ammonia, or nitrate salts. With the great diversity of organic substances, accordingly fermentation processes and their end products quantitatively and qualitatively are very different.

It is sufficient to point out here that all those fermentations by which fungi withdraw their oxygen requirements from the organic substance are gathered together under the collective term putrefaction (Fäulniss).

If the rotting body was rich in proteinaceous substances, then fetid gases such as hydrogen sulfide and so forth develop, while from bodies poor in such materials, as for example woody bodies, only few such gases arise. The putrefaction of protein deficient bodies often is labeled rot (Vermoderung) or humification. The end products of putrefaction are always deficient in or entirely free of oxygen.

All bodies rot in their interiors where the air finds no free access.

All processes of fermentation, in which oxygen of the air has easy entrance, are called decay (Verwesung).

Fungi withdraw their oxygen from the air, and should generate it in the substances which are attacked in the decomposition.

A direct oxidation, however, must certainly not be ruled out. The end products of decay are very rich in oxygen: carbonic acid, water, and nitrates. All bodies which are exposed to the air rot at least externally, while they often simultaneously putrify within.

The parasitic fungi or parasites, which are known as causal organisms of numerous diseases of plants and animals, present a much broader, many sided interest. The fact that among those ignorant of fungi, mostly only those fungi that confront us all as the most common causal organisms of putrifaction or decay are generally known and have been observed, that their known life cycle is transferred unknowingly to all fungi, explains the fact that, at least among forestry people, there still today exists a deep rooted bias against the results of scientific investigation in the general area of the teaching of plant diseases.

The less one has concerned himself with observing and investigating plant diseases, as a rule with the greater certainty he defends the opinion that infectious diseases have completely other causes, that the fungi appearing therewith arise only subsequently, that they are purely of secondary nature. The answer to the question of what then is the true cause of the disease as a rule surely will be absent, if not from perplexity the discussion is turned to the weather or if unknown poisonous substances lying in the soil and so forth must be held out.

No one will doubt that the weather is able to exert an unfavorable influence on the development of plants, that the frost kills the foliage, and thereby can considerably damage the growth of plants, that also possibly sudden temperature changes are detrimental to many plants. Poor, barren or too damp a soil, thin soil, underlying hard pan and so forth can impair the growth of plants. The diseases which occur as a result of such influences of inorganic nature almost always show an entirely different character from those diseases which are caused by parasitic fungi. The latter almost always are recognized at first sight as contagious diseases; they begin on one point on the plant, from which they gradually spread out further; they occur first on individual plants, which to a certain degree form the center for further infection in all directions. On the contrary, unfavorable weather influences strike an entire stand more or less in all parts simultaneously and suddenly; unfavorable soil influences have as their result slow stunted growth from youth on or a gradual decline of the stand, but can never explain the sudden killing of vigorously growing trees or the spread of disease from one point over a larger area in the course of years.

Among the diseases of forest trees which have been more closely investigated by me till now, not one can be named which justifies the opinion that the effectiveness of the associated parasitic fungi is limited or promoted by a certain pathological predisposition of the trees. Of itself, the possibility of such preliminary conditions for the effectiveness of a parasite is excluded in those diseases which I was able to produce artificially, by infection, on every desired plant.

However, those fungus diseases which I have not been able to produce through artificial infection so far can be explained by the fact that the development of the fungus in all of its forms, the duration of spore dormancy, germinability, and conditions for germination are still not sufficiently clarified. The type of behavior, the gradual progression of the disease, and other factors, however, permit no doubt to occur about the parasitic character of the observed fungi.

The fungi always get into the interior of the host plant in this manner: the spores germinate externally and the germ tube penetrates into the interior through the stomates or the epidermis of the leaves and young bark, or into the roots, or into surface wounds of the stem, to develop there the mycelium. Only *Agaricus melleus* forms an exception; its mycelium grows in the soil from one plant to another and penetrates into their roots.

The spread and effect of fungus mycelium in the interior of the host plants present the greatest diversity.

According to the part of a plant which is attacked, the following fungi are described as parasitic to the roots: *Agaricus melleus, Trametes radiciperda;* of the woody tissue: *Trametes Pini;* of the bark and the phloem: *Peridermium Pini* and *Caeoma pinitorquum*; finally, of the leaves: *Peridermium Pini, Caeoma Laricis, Hypoderma macrosporum, H. nervisequium,* and *Melampsora salicina.*

According to the permanence of the mycelium they can be divided into those of at the most one year duration: *Caeoma Laricis, Melampsora salicina;* those with two to three year duration: *Hypoderma macrosporum, H. nervisequium,* and *Peridermium Pini;* those with several years' duration: *Agaricus melleus, Trametes Pini, T. radiciperda, Peridermium Pini, Caeoma pinitorquum.*

The effect of the mycelium on the cells and on the cell contents of the tissues in which it grows likewise is different. Just as in the case of the saprophytes where the breakdown of organic substances into simpler chemical compounds is explained by the fact that after the removal of certain portions by the fungi, the components of the organic substances which remain behind must combine into new compounds, so also the effects of parasites on their host plants is explained, above all, by the withdrawal of nutrient materials which the mycelium needs for its growth.

The mycelium of *Peridermium Pini* exerts only a very slight detrimental influence on the life of the cell, as is seen by the fact that severely attacked needles into the parenchyma of which the mycelium has spread, do not die off after sporulation of the aecia, but, on the contrary, remain green up to the immediate vicinity of the fruit bodies.

The most noteworthy effect of this mycelium exists in the transformation of the starch to turpentine, apparently by withdrawal of oxygen, by which a saturation of the tissue with resin is brought about. The hyphal threads of *Agaricus melleus,* growing into the resin canals, destroy the thin-walled and starch-filled tissue in their margin so that large cavities arise in the wood where resin canals previously occurred.

The extremely abundant resin flux on the root stock, the formation of resin boils in the bark, the formation of resin cavities in the annual ring of the upper half of the root stock which is formed during the diseased condition, can be explained only by the transformation of starch and cellulose to turpentine under the influence of the fungus hyphae. Further proof of the conversion of starch into turpentine is given by the occurrence of the latter in the medullary ray cells of barkshake pine wood (Plate III, Fig. 18). The destruction of the walls of the wood fibers by *Trametes Pini* is essentially hastened by the numerous holes which the mycelial branches bore through the walls. From the borders of these holes outwards the walls are especially severely attacked and, indeed, chiefly the middle so-called cellulose layer, while the inner layer of the cell wall offers the longest resistance to destruction and disintegration (Plate III, Fig. 18).

The mycelium of *Hypoderma* is immediately lethal to the leaf parenchyma of spruce and fir (Plate VI, Fig. 5), which shrivel up as soon as the mycelial threads come in contact with them in the intercellular spaces. The same is true for the gelatinous sheath at the end of the rhizomorph which, pushing itself through between the living phloem cells, immediately results in a browning of the phloem tissue in contact with it.

Agaricus (Armillaria) melleus L.

(Plate I and II)

The Honey Fungus (Hallimasch)

Causal Organism of Resin Flux (Harzsticken, Harzüberfülle), Root Rot, or Ground Canker (Erdkrebs) of Coniferous Trees

The biology of fungi of the genera *Agaricus, Trametes, Polyporus,* etc., has been ignored to a remarkable degree by mycologists until now even although as the following work can prove, the study of these can lead to the most interesting results. The fungus, whose description forms the content of this first section, is the causal organism of a disease which is numbered among the most widespread and most destructive in the coniferous forests of Germany and well beyond its boundaries. The more or less sudden dying of Scots pine *(Pinus sylvestris)*, Weymouth pine (*P. strobus*), black pine (*P. niger*), dwarf mountain pine (*P. mugho*), cluster pine (*P. pinaster*), spruce, fir, and larch in different age classes from about five years old up to, at least in the case of Scots pine, over 100 years old, characterized by an abundant resin secretion on the root stock and larger roots, further by a white fungus mycelium which occurs under the bark of the roots and lower part of the stem, from which dark brown fungus strands similar to fiberous roots grow out through the soil, under the terms listed above is a phenomenon known indeed to all forest workers in coniferous forests.

In spite of this, only very few reports about this disease occur in the forestry literature.

1. In the Verhandlung des Hils-Solling Forstvereins of 1861 is mentioned "the cause of the sudden dying of transplanted young spruce." It was reported that a similar disease was observed on Scots pine and Austrian pine. The illness was sought to be explained by a "sap weeping" (Saftüberfüllung) due to "rupturing (Sprengung) of the crowded together roots of group-planted seedlings" and to other causes.

2. In the Verhandlungen des Harzer Forstvereins of 1864 occurs a report by my father, Theodor Hartig, on the "Resin flux of spruce":

"Earlier in the group-planting of spruce seedlings in the Harz, there has been observed here and there a dying of the spruce seedlings, after previous

[12]

bursting of the bark of the root stock just above the soil, and an abundant resin discharge from the wound crack, by which the earth around the roots was almost cemented. The disease was related to a hindering of the movement of the resinous sap through mutual root pressure, which occurs frequently in spruce seedlings."

"I, myself, was of this opinion until I had opportunity to observe this disease last year in a seven-year-old single tree planting in the Heinrichswinkel Forestry Compartment of the Braunlage District."

"The investigation gave the following:

"As the immediate cause of the disease there appears indeed to the naked eye an excessive secretion of resinous sap; the organs which are destined to the manufacture of this occur in unusual size and number in diseased plants. The large resin vessels of the green bark reach in part a diameter three to four times larger than customary. One recognizes this externally in the hemispherical pustules of the bark, which in part are split open and from which the turpentine has flowed out externally. The resin canals, which occur in smaller numbers and separated in the wood of healthy spruce, are so numerous in the youngest annual ring of diseased spruce that they form a closely crowded-together row. It is not uncommon that this row of resin canals blend into one another and form a common cavity up to the size of a bean, which then is completely filled with resin."

"It is well to consider here that the disease consists not only of an abundant secretion of this resinous sap but also the organs for this secretion are formed in an excessive number, the cause for the disease also lies deeper than in the increased activity of the individual secretion organs."

"The outermost annual ring, which is filled with resin ducts, always remains uncustomarily narrow. There occurs in the majority of the cases an unusually broad annual ring with normal number and size of the resin organs immediately before, from which we must conclude that the primary causes of the disease are not unfavorable external conditions and influences but, on the contrary, excessive support and extraordinary enhancement."

"The disease makes short work, since already in the year of its occurrence the plant which is attacked by it dies. There has appeared to me at least until now, also not from earlier collected and preserved material, no case in which more than the outermost annual ring shows uncustomary increase of the resin canals. The disease therefore is well to be feared, if it should occur in greater distribution and more frequently, which until now has not been the case."

"In the xylem as well as in the phloem the resin flow is accompanied by a fungus formation which perhaps is unique in its mode of development."

"On those spots of the xylem as well as on the phloem, which lie next to the resin secretion, there begins a progressive transformation of the xylem and phloem fibers into a fungus distinguished in that it appears not in isolated fibers, as is usual under similar conditions, but immediately in a completely continuous cell tissue, similar in form and arrangement to the parenchymatous cell tissue of the pith or the bark of higher plants."

"This milk white tissue of cells, without any permanent contents of the individual cells and without spore formation, spreads out radially between the fiber layers of the xylem and phloem, similar in outward appearances for the most part to *Ozonium candidum* (*Himantia candida* Pers.), but distinguishes itself from this, however, by the primordial anastomosis of the threads into a continuous tissue of cells, and also in that the fungus never occurs externally but also, like the night fibers (Nachtfasern) of the red and white rot always occurs only between the xylem and phloem layers. Viewed from the rules of nomenclature of fungi, the name *Ozonium parenchymaticum* would be characteristic."

"In any case, this fungus is not the cause but certainly a general consequence of resin flux."

3. Willkomm wrote in "Die mikroskopischen Feinde des Waldes," Volume 11, Page 187. I have also observed in the Harz that with young larch also death can result from resin flux from the root collar—the same phenomenon which has been observed for a long time previously on young spruce and Scots pine, but still is not clarified, furthermore, is accompanied by a still not closely investigated fungus growing under the bark.

4. Theodor Hartig, in the "Kritische Blätter für Forst- und Jagdwissenschaft," 51(I), 1868, under the appendix of an illustration once again gives a description of the disease which essentially agrees with the report of Harz Forstverein. The fungus mycelium, however, was identified correctly as *Rhizomorpha subcorticalis* and *R. subterranea.* The earlier expressed opinion of the secondary character of the fungus was retained with great certainty, especially on the basis of the fact that no one would subscribe to the hypothesis that the overproduction of turpentine vessels could be brought about by fungi.

5. Finally there occurs in the "Zeitschrift für Forst- und Jagdwesen" von Danckelmann, 1870, Volume 2, page 359 forward, a short report by me on the "Auftreten der Rhizomorpha in Nadelholz-Culturen," in which was discussed especially the distribution of the disease, its progress according to external symptoms and so forth. The opinion of Theodor Hartig was already drawn into doubt.

In the meantime I have been successful in obtaining a completely clear insight as to the nature of this disease and I publish in the following the results of my investigations and observations.

THE MYCELIUM OF AGARICUS MELLEUS

The mycelium of *Agaricus melleus,* which is known as the causal organism of the disease, is differentiated essentially from that of most other fungi in that it occurs not only in a simple-threaded form but also in the form of cord- or ribbon-shaped fungus bodies.

Rhizomorpha fragilis Roth, with its two chief forms *R. subterranea* and *R. subcorticalis,* belongs as the mycelial body to *Agaricus melleus* and no observations exist which could be of value as more certain proof that *Rhizomorpha fragilis* also produces still other fructification organs. It seems unlikely to me also that still other fungus species produce mycelial bodies which are so similar to *Rhizomorpha fragilis* that confusion could easily occur.

The simple-threaded mycelial form can be discussed only after the structure of *Rhizomorpha fragilis* has been described.

The rhizomorphs, which bear the most peculiar character, which we subsequently learn to recognize, are distributed everywhere in dead roots of hardwoods and conifers, between the wood and bark of dying trees, in water pipes, in old bridges, etc.

Schmitz, who discussed the different forms in his excellent description of *Rhizomorpha fragilis* in Linnaea, 1843, is of the opinion: "That all rhizomorphs develop only at the ground at the bases of tree trunks, also preferably in forest soils or between wood and bark, when sufficient moisture is present here, that all rhizomorphs which occur in mines (not only coal mines but also ore mines), in deep pipes and canals, occur there not in their original habitat but in a secondary habitat. This theory has its proof in the fact that all rhizomorphs which have been dried for a long time and are dead, very easily come to life again with moisture and can grow, and that therefore old rotting wood, in which the remains of rhizomorphs are enclosed, under certain conditions can form new shoots. Because the timbers in mines consist chiefly of oak and beech wood and only here do the rhizomorphs gain their nutrition, thus must we conclude with certainty that rhizomorphs growing in this wood in underground regions are only an accidental phenomenon and likewise have been displaced here from the forest, etc., etc."

This opinion fully agrees with mine; I notice in regard to different forms which the rhizomorphs take under different conditions for growth, that those rhizomorphs which develop in entirely free air as well as those in the soil or externally on the roots take on the form of round, frequently branched, dark-brown-colored cords, easily confused with fibrous roots, the thickness of which varies between 0.5 and 3mm diameter. Such cords we see in Plate I, Fig. 1, 2, and 3, and Plate II, Fig. 1a, b, c, 5b, 6, 11, 12, etc. The cord in Plate I, Fig. 3, developed between the bark and wood of an already dead tree with only very moderate pressure of the bark; it was somewhat compacted together and has two rows of lateral branches.

This type forms almost a transition between the previously described form designated by the name *Rhizomorpha subterranea* and the second form *R. subcorticalis,* which develops as soon as the rhizomorph grows into the living phloem of coniferous wood or because of the pressure of narrow splits and cracks in the interior of dead trees or in cracks in stone cannot attain an unhindered formation. This second form produces very thin broad ribbons or mycelial skins which spread out fan shaped, as those illustrated in Plate I, Fig. 4, 5, 20, 26; Plate II, Fig. 2, 5a, and 11a.

The transition of the two forms into one another is demonstrated several times in the designated illustrations.

The edge of this mycelial body which spreads out flatly is deeply incised (Plate I, Fig. 4), or deeply indented (Plate I, Fig. 5), or there arise cords which unravel themselves into fine threads (Plate I, Fig. 5c), or round cords of *Rhizomorpha subterranea* (Plate I, Fig. 5, Fig. 4a, etc.).

The brown rind of the second form usually grows in such close relationship to the wood and bark of the tree that with the shedding of the latter the rhizomorph is almost always torn apart in the middle and thereby the light core of the mycelial body appears.

The structure and development of *Rhizomorpha fragilis* have been described by de Bary* in such an explicit manner that I can follow verbatim his results, in order to touch partly upon my own continuing investigations in part and to point out the illustrations:

"The fully developed cord consists of a dark brown, paper thick, brittle, mostly smooth rind, which surrounds a white, fine threaded tough core. The rind (Plate II, Fig. 17, 18, 19) is formed of at least 12-15 layers of rows of cells (hyphae), which occur parallel to the length of the strand and which grow together with one another compactly and without intercellular spaces. The hyphae of the outermost layers (Fig. 17f, 18, 19a) are composed of narrower and thicker walled cells than those of the interior, the individual cells 2-4 times longer than wide, possessing a solid brown, clearly layered wall, often polygonal in cross-section. The lateral adhesion of the walls is often so strong that in thin cross-sections they appear to form a homogeneous mass; especially with treatment with KOH (Fig. 19), however, clear borderlines appear, which separate the apparent homogeneous mass into a number of walls which correspond to the number of existing cell lumina."

Only the freely growing rhizomorph cords possess such a rind, while the ribbon or flatly spreading rhizomorphs, which develop in the phloem of trees, almost always have a very thin rind which is restricted to a few hyphal layers, of which the outermost does not become brown (Plate I, Fig. 18). Often, indeed, the rind consists of only two or three cell layers, which form a thin skin firmly growing together with the xylem and phloem.

On the other hand in very thin mycelial skins of *Rhizomorpha subcorticalis* the core is often completely lacking or is composed of only a few hyphae. The rhizomorph then consists only of a very thin-skinned rind.

I have observed the absence of the core most frequently in the downy branches which arise from the margin of larger mycelial bodies (Plate I, Fig. 5c). The distal points consist then only of rind hyphae (Plate I, Fig. 9) from which individual thin strands, consisting of a few isolated hyphae, (Plate I, Fig. 6, 7, 8) separate and form a type of threadlike mycelium, characterized by cells with clamp connections and the frequent appearance of a granular surface (Plate I, Fig. 8).

"On the inner side of the rind there is usually a thin tissue layer, or at times even surpassing the rind itself in thickness, which is composed of light brown and on the cross-section very irregularly tightly woven and thickly compacted threads, which on the one hand arise from the interior rind elements and on the other side are transformed gradually into the colorless hyphae of the core."

---

*Morphologie und Physiologie der Pilze etc., p. 22 ff., Leipzig. 1868.

"This consists usually of thin, about 1/560 mm thick threads which are intertwined at acute angles and which run lengthwise (Plate I, Fig. 16 and 17 g.g.). The wall itself is rather firm; cross walls and branches occur only rarely in growing samples. Between these hyphae are distributed thin-walled colorless cells shaped like cylindrical tubes, which occur often at fairly regular spacing from one another, up to ten times as thick as the mentioned threads (Plate I, Fig. 17b). In old samples they are often difficult to find. The interstitial spaces of the core tissue contain air."

I have never observed a layer of tissue different from the core lying on the inner side of the rind.

The outer rind, which consists of narrow lumened thick walled hyphae (Plate I, Fig. 16, 17f, 18, 19a), is transformed gradually into the inner rind which consists of large celled thin walled hyphae (Plate I, Fig. 16, 17g, 18, 19b). The core always immediately follows provided that the formation of side branches or rather of fruit bodies does not explain the presence of a pseudoparenchyma on the inner side of the rind (Plate I, Fig. 17m; Plate II, Fig. 14m).

"Old very thick samples of the cylindrical form (*R. subterranea*) often have an uneven wrinkled rind, in which, perhaps through later growth, the number of cell layers has increased greatly with an irregular arrangement. In the interior of such samples I have often found, although not always, a brown zone which is concentric with the rind, but separated from it by a narrow layer of usual core tissue and in turn surrounding a strand of the latter. This zone consists of threads which are brown skinned and very strongly matted together with one another, which moreover are similar to the usual elements of the pith, and also are continuously transformed into the latter."

Eschweiler's description of the structure of rhizomorphs is probably without doubt based on the investigation of such samples."

The assumption that later growth leads to an increase in the number of hyphae in the rind seems to me very unlikely and must at least be proven. After the formation and browning of the rind I do not consider that growth in thickness of the strand could occur without bursting the rind. I have never found rhizomorph strands with a second browned inner zone.

If one cultures thick rhizomorphs in a moist room, after about 8 days new branches develop. First, at arbitrary points on the surface small (about 0.5-1mm large) white flecks appear, usually gathered together several to a group. They consist of branched, tortuous hyphae gathered together into a group, whose free ends appear colorless and thin walled, whose bases on the contrary possess a more robust brown wall (Plate II, Fig. 7, 12, 14n, 9). They arise from the inner cells of the rind as branches, growing from here outwards are united into a cylindrical cord perpendicular to the outside, break out through the outermost layer of the rind and then grow radiate to one another. These little tufts are the precursors of the branches; with the appearance of the latter the tufts disappear and one finds only their disintegrated remains. In the same spots where the little tufts arise on the outside, there simultaneously begins a new formation on the inner surface of the rind. There arises here a thick

[17]

parenchyma-like tissue consisting of moderately wide, irregularly long, very thin walled, hyaline cells, which in part are placed quite irregularly, partly in rows running perpendicular to the upper surface (Plate I, Fig. 17m; Plate II, Fig. 7, 14m). As far as I could determine, because of the great thinness and strong binding together of its cells, this tissue always originates from the buddings which arise from the innermost cells of the rind. Their elements increase quickly and considerably, they lie tightly on and between the peripheral hyphae of the core, often forcing many of these out from their exact longitudinal orientation to such an extent that they pass sinuously through the newly formed tissue and separate the rind from the core with which they are always strongly connected. Immediately under the little tufts new growth often is especially active; here the rind now and then lifts up toward the outside in the form of a cylinder stuffed with colorless young tissue, which supports the little clump like a short stipe. After a few days the primordia of one or more branches emanates from this newly formed tissue. A portion of this tissue grows out in the form of a conical wart with the point turned against the old rind, breaks through this and appears from the crack as a white, cone-shaped body, a branch primordium.

Up to this point the description of the origin of branches of the rhizomorph agrees essentially with the origin of young fruit bodies which I was successful in observing.

In particular regarding the origin of the tufts of hair, they arise also from the outer surface of the rind (Plate II, Fig. 7 and 14nn), because very often not the slightest breaking of the rind under the tufts of hairs can be observed. Also in the proximity of the spots in which the rind is broken through, one often sees individual hairs arising without doubt from the surface of the rind. The surface of the hairs shows characteristic granules (Plate II, Fig. 9), which, when treated with potassium hydroxide solution, in part became detached. Sometimes, although rarely, similar hyphae arise in the tissues of the emerging young fruit body, as Fig. 10n shows. I observed a very conspicuous formation at the base of such a hair. From the basal cells arose a sterigma-like growth; on its top was found a dark brown colored round cell (Plate II, Fig. 10o). The formation of pseudoparenchyma under the tufts of hair is illustrated in Plate I, Fig. 17 m. From the innermost rind cells arise numerous hyphae, which are distinguished from the medulary threads even in the beginning by a greater thickness, above all, however, in that the double layers of the cell wall, frequent branching, and numerous crosswalls are conspicuously apparent. Not far from their point of origin they take on considerable dimension and form the pseudoparenchyma which is characteristic for the tissue of the fruit bodies (Plate I, Fig. 17m; Plate II, Fig. 10m).

The surface cells at the base of the branch rudiments immediately have brown walls; they exhibit a rind with an uneven surface formed from round or long cells, which remains firmly grown together with the inner rind of the old stem. In the middle of the branch primordium the cells stretch themselves out in the direction of the longitudinal axis; with this elongation their row-like arrangement always appears more clearly; they remain colorless and thin-walled in firm connection with the irregular pseudoparenchyma under the older rind. Near the point of the branch primordium the rows of cells (hyphae) converge as they combine (Plate I, Fig. 10). The axillary ones progress straight, the peripheral ones bend bowlike towards these; thus arises a more or less rounded off conical end, which is considered to be the growing point of the young branch. The thickness of the hyphae and the length of the individual cells decrease steadily towards the growing point; the former reaches here about 1/450 mm. All hyphae which converge behind the growing point lie very close of one another, the peripheral ones already quite without cavities; often very early air-containing interstitial spaces occur between the axillary ones. The entire surface of the described branch primordium is overlain by a loose reticulum of branched hyphae, mostly 1/450 mm thick, here and there thicker, which arise as branches from the surface cells of the branch primordium (Plate I, Fig. 10, 11e). Between these threads lies a homogeneous colorless jelly which swells strongly in water, whereby the surface of the branch becomes slippery. This thickness of the young branches in my cultured samples averages 1 mm."

The characteristic jelly of the growing rhizomorph end appears never to be absent; however, sometimes it elongates, which gave occasion to draw those rhizomorphs which develop between the wood and phloem of a young suddenly killed cherry tree (Plate I, Fig. 26).

The jelly (c) which has been secreted by several ribbon-shaped rhizomorphs (d) forms a layer about 5 mm thick and is a brownish color. In all parts this is penetrated by a reticulum of many-branched hyphae, as is shown in Plate I, Fig. 11e, and which dissolves almost completely on treatment with potassium hydroxide so that the embedded hyphae become completely free.

The hyphae in the jelly are abundantly branched, numerously septate, and show occasionally characteristic ball- or pear-shaped swellings (Plate I, Fig. 12), whose significance is not clear, since the supposition that they are gametes (Befruchtungskugeln) has too little justification. That these hyphae arise from the rind hyphae of the rhizomorph can scarcely be doubted, that they arise directly under the growing point follows from the orientation which is indicated in Fig. 10, which at (d) is predominantly radiate; 1 mm behind the point, on the contrary, it is already predominantly parallel to the longitudinal axis of the strand. Now and then these hyphae only 5 mm behind the point show a highly remarkable cell wall formation (Plate I, Fig. 13).

The original overall similar, rather thick cell wall here is broken into longer or shorter fragments, which can be explained only by a stretching of the hypha after the cell wall has lost the ability to expand.

The torn cell wall pieces are bound together by the stretched out plasma membrane (Innenschlauch). An analogous cell wall formation must occur only extremely rarely and has never been observed by me till now.

After the appearance of the old rind, the branch grows in length by a continuous growth of all hyphal ends which are united into the growing point. This permanently retains its original condition; the closely compressed hyphae in it are always short segmented, rich in protoplasm, and, as far as can be determined, are all similar to one another.

Close behind the growing point begins the stretching of the cells and the separation of the tissue into an axial portion, which I will call the primary core, and a peripheral one, the rind.

The former one forms a white, thin cylinder (Plate I, Fig. 10, 14b) consisting of hyphae whose cylindrical, thin-walled, and for the most part hyaline, liquid containing cells rather close behind the growing point attain a thickness of 1/75 to 1/50 mm with 2-8 times greater length. Air-containing interstitial spaces occur between the hyphae.

Now and then the cells of adjacent hyphae anastomose in pairs into the form of an H. Towards the periphery, the hyphae of the primary core become thinner and are transformed gradually into those of the rind (Fig. 10, 11, 14 b-c).

This consists of narrow hyphae which lie close to one another, of which the outermost in the young rind are held together in a loose connection with the remaining ones only by a thin colorless jelly (Fig. 10, 11, 14f). From these arise the threads of the above mentioned loose and jelly surrounded tissue, which always surrounds the young rhizomorph branch, and which is renewed continuously from the growing point."

The previously described structure of the round rhizomorph apex is fully in agreement with the structure of the flattened *Rhizomorpha subcorticalis,* self evidently with the modifications obviously caused by the form. A section made at right angle to the broadside of the rhizomorph (Plate I, Fig. 4b) shows exactly (Fig. 10) the described picture.

The branches which are pointed wedge-shaped or which turn into fine pointed hairs (Fig. 6, 7, 8) completely lack the primary core; moreover, they consist only of extremely fine hyphae corresponding fully to the peripheral rind hyphae of the previous form (Fig. 11f). If numerous rind hyphae are joined together (Fig. 9), then one can see quite clearly the boundary of jelly which holds them together, but if the latter apexes are formed only from individual hyphae, then only the cementing of these infers the presence of a jelly.

"For a distance of at least 2-3 mm behind the growing point the young branch is always quite colorless. Further back the rind begins to take on a more and more intense beautiful brown color; at first the walls of perhaps six hyphal layers lying under the surface become colored (Plate I, Fig. 16c and 18b), and simultaneously become thickened; an external somewhat similarly thick layer shows the coloration and thickening of the walls more slowly and later. Simultaneously with the latter, the loose reticulum of hyphae of the surface takes on a brown color, the jelly is firmer and less capable of swelling, and likewise is brown. The latter parts develop thus into the external narrow-celled rind tissue, the former colored layers into the inner rind tissue. In a few cases I saw numerous colorless branches bud out from the hyphae of the rind which was just becoming brown, which together formed a thick coating of equal-lengthed hairs, extending at right angles and bound together by jelly, which later disappeared. In most cases these formations did not occur."

As regards the browning of the rind, at least in the case of the form *subcorticalis,* which develops in living phloem, this deviates from the previously described manner insofar that very often either the browning does not occur at all or only in the inner rind or only in the outer rind. The opinion that the loose-threaded reticulum of the surface with its surrounding jelly becomes the outer rind is incorrect, if the hyphae are common with the former (Fig. 10 and 11e). The latter apparently very soon lose the ability to increase in size; they divide usually very irregularly on the outer surface of the rhizomorph which grows significantly in size by cell growth, and only tufts of hair (Fig. 13) can be seen as small brown pustules at x (Fig. 4) perhaps 1 cm behind the point. At the beginning of the browning of the rind, these hyphae distributed irregularly in the jelly for the main part must have undergone their destruction. The composition of the outer rind from hyphae which run parallel to the axis of the strand (Plate I, Fig. 17f, 18, 19a) speaks to the fact that they arise from the outermost rind hyphae designated f in Fig. 11 and 14.

From these also arises the highly important, thread-like mycelium of the fungus, whose significance de Bary has not recognized. The so-called hairs observed by him in several cases arise in every *Rhizomorpha subcorticalis* which develops between the wood and bark as the characteristic nutrient absorbing organs of the fungus mycelium. The roundish strands of *R. subterranea* which creep around in the soil, in the air, and on the surface of ·wood, and which only spread the fungus to a certain degree, perhaps may absorb barely sufficient amounts of nutrients.

Directly behind the apex of a young *R. subcorticalis* arise the perpendicularly extending mycelial threads of the rind hyphae (Fig. 10, 11, 14, and 16h), which grow through the jelly layer and then branch out repeatedly.

These hyphae penetrate into the phloem and the bark, passing preferably through the medullary rays to the interior of the xylem (Fig. 14, 15, 16).

In coniferous woods they seek out above all the resin canals, growing in these rapidly upward, considerably higher up within the tree than *R. subcorticalis* has developed in the phloem.

The cell tissues containing starch in the vicinity of the canals are destroyed by the fungus hyphae so that finally a large hole indicates the spot where earlier a canal has existed (Plate I, Fig. 23, 24, 25). We will later see that not only the resin secretion on the stump but even the formation of numerous resin holes in the bark and newly formed xylem can be explained by the development of this thread-like mycelium in the resin canals of conifer wood during the year of the disease.

This thread-like mycelium plays an extremely important role, above all where rhizomorphs develop on dead hardwoods and conifers; likewise, on wood used in buildings. The hyphae penetrate into the wood, growing especially in great numbers in the vessels (Holzröhren), also in other organs, and cause thereby a quick destruction of the wood. Especially from the lumina of the medullary rays and from the openings of the vessels, the hyphae often appear in the form of rust-colored tufts, which in their formation are completely in agreement with the tufts of hairs described on the rhizomorph strands (Plate II, Fig. 12, 14, and 9).

"The innermost brown rind cells and the outermost layers of the primary core, already before the beginning of the browning, expand considerably in thickness and width; the axillary rows of the primary core show this expansion in the designated directions to a lesser extent; they stretch considerably only in length (up to 20 times the cross-sectional diameter). The branch thereby increases in circumference, the axillary hyphae are torn apart from one another, the air-containing holes between them widen significantly. The primary core (Fig. 10 and 14b), as is clear without further adieu, has a structure totally different from the core of the mature rhizomorph. Where the browning of the rind begins, there begins now the formation of a definite core. All colorless cells occurring within the rind as well as the innermost rind cells themselves now put forth partly from their side walls, and partly also from their ends, thin, branched hyphae (Fig. 14g), which in the beginning appear thin-walled with clear cross-walls and turbid plasma contents, which rapidly lengthen and take on the character of mature medullary hyphae. The growth of these threads is parallel to the longitudinal axis of the branch, directed partly towards the point and partly towards the base; they push themselves everywhere between the cells of the primary core, increasing and lengthening, so that very soon they compress those and force them together, forming the main mass of the tissues which are surrounded by the rind: a thick, fine-threaded strand in which the cells of the primary tissues lie scattered about (Fig. 17b, g) as broad, thin-walled tubes, often detected only with difficulty. Often the core only 1 to 2 cm behind the growing point has this structure. From the bases of the young branch the medullary hyphae pass between those of the old main stem and entwine themselves with it so finally the core of the latter carries on continuously into that of the branch."

There remains for me only to note that a fresh *R. subcorticalis* is distinguished by a very pleasant odor completely corresponding to that of the fruitbody of *Agaricus melleus,* as well as by a characteristic phosphorescence which can be detected in the darkness at a great distance.

THE FRUIT BODIES AND REPRODUCTIVE ORGANS

For a long time the search for the fructification organs of *Rhizomorpha fragilis* has concerned botanists and has led to the most varied hypotheses, indeed, to be regarded as settled at the present time.

De Bary, who in his *Morphologie und Physiologie der Pilze* gives a review of these works up to 1865, comes to the conclusion that, "according to present observations it is still not possible to say with certainty to which fungus species *R. fragilis* belongs."

Since the earlier opinions to the present can serve only an historical interest, thus indeed a short review of these is sufficient.

DeCandolle, Eschweiler, and Acharius believed they had found a pyrenomycete fruiting body on the rhizomorph which, however, later was considered as an outgrowth and branch primordium.

Fries considered just recently the rhizomorph as a perithecia-bearing genus of pyrenomycetes.

Otth found situated on the rind of *Rhizomorpha subcorticalis* small, black, bristle-thick, 1-4 mm long bodies which bore on their apices one of the species of *Stilbum* or a *Graphium*-like fructification. The opinions of botanists about these bodies found by Otth are different; by now the opinion of Montagne and Cesati, which holds these as a parasitic *Stilbum* species, is probably correct.

Palisot de Beauvais found a rhizomorph whose branchings united into an irregularly spreading *Xylostroma* whose margin developed without doubt into a fruiting *Polyporus*.

Caspary concludes from the occurrence of fruit bodies of *Trametes pini* on Scots pine, which also showed *Rhizomorpha fragilis,* that, indeed, the fruit bodies are of the latter.

Caspary asserts also to have found *Polyporus* species and *Agaricus ostreatus* in connection with *Rhizomorpha*.

Tulasne observed fruiting *Polyporus cuticularis* and *P. alneus* developing from *R. subcorticalis* and so forth.

Haller, Bail, and Lasch claim to have observed fruit-body bearing *Xylaria hypoxylon* situated on *Rhizomorpha*. Bail has corrected his results, in which he stated that he was unable to demonstrate an inner connection of *Xylaria* with *Rhizomorpha*.

Fuckel claims to have found without doubt, as recently as 1870, the formation of fruiting bodies of *Rhizomorpha* in the form of perfect, carbonaceous perithecia about 1 millimeter high and 0.5 millimeter broad, rounded off on the under side, extending on the upper side to a glossy black, ostiolate conical snout, which at their bases were buried only slightly in the stroma.

In the interior they contain free elliptical, somewhat scalene, smooth brown spores 16-20 microns long and 8 microns broad, filled mostly with 8 round hyaline sporidiols (Sporidiolen).

In regard to the results of my own investigations I must doubt that the characteristic and true reproductive organs of *Rhizomorpha fragilis* are the perithecia of Fuckel. On the contrary, I believe to be able to prove that the illustrated fruit bodies of *Agaricus melleus* (Plate II) are the sole universal and regularly forming fruiting bodies of *Rhizomorpha fragilis,* and I am fully in agreement with the opinion of de Bary in which he says:

"It is, in fact, very doubtful that these always consistent characteristics (in the form of the structure of *R. fragilis*) should belong to quite different fungus species; at least it must be doubted until definitive proof of this is presented. That there are rhizomorph-like mycelial strands deviating in structure from the typical rhizomorph is proven by the observation of Tulasne, who found a form of *Xylaria hypoxylon* arising from a long, black rhizomorph strand. The white middle substance of the latter consisted of thick fibers, whose walls were thickened up to a point of disappearance of the lumen; it is surrounded by a black rind of round laminar or spherical polyhedral cells, etc."

The fruit bodies of *Agaricus (Armillaria) melleus* develop at the beginning of October partly from the root stocks of such trees under whose bark *R. subcorticalis* occurs, as soon as these can come forth externally through bark cracks and the like, partly also on the apices of rhizomorph strands in the soil.

Plate II, Fig. 4 shows a young Scots pine suffering the disease with numerous tuft-like fruiting bodies breaking forth from bark cracks (aa) whose connection with the edge of *R. subcorticalis* is easily proven and in Fig. 5 is discerned by uncovering this after removing the bark. Individual cords of *R. subterranea* growing forth from the bark bear at the tip more solitary fruit bodies (bb).

The position of the fruit bodies on the strands of *R. subterranea* is very variable. Either they sit individually directly on or somewhat behind the tip of the strand (Plate II, Fig. 6a, b, 8, and so forth) or simultaneously several fruit bodies arise on the tip of the strand (Fig. 6c, 11, 1c, and so forth). Of these the majority shrivel up (Fig. 1c, 27, 28, 29, and 30) or they subsequently grow into a single fruit body (Fig. 31).

Often one sees, however, numerous fruit bodies breaking forth laterally from the rind of the strand. In Fig. 1b, numerous fruit bodies are breaking forth on both forks of a strand, of which only the uppermost are going to develop fully while those below become smaller closer to the bifurcation until they apparently transform into tufts of hair, which designate the first beginnings of branch and fruit body formation. These points of forking are especially illustrated in natural size in Fig. 12.

The rudimentary fruit bodies remain mostly light, often without clear cap formation; sometimes they appear only as dark-brown-colored tufts near the attachment place of the developed fruit body (Fig. 29 and 30).

The formation of the fruit body initially is in complete agreement with the origin of a side branch, as we have learned on pages 17 and 18. Under a tuft of hair there arises on the inner side of the rind a parenchyma-like tissue (Plate I, Fig. 17m) which breaks through the rind and appears as a small, light-colored button.

In Fig. 7 and 14 one sees the pseudoparenchyma (m) appearing externally after the rind has been split under the tuft. Fig. 6a shows in natural size the first undoubtable meristem of a fruit body which is magnified larger in Fig. 7. The pseudoparenchyma (mm) forms the stroma itself and is depicted in Fig. 10m consisting of irregularly interwoven, large-celled hyphae. On the outer circumference of the fruit body, the hyphae become brown and show highly varied, differently shaped end members (Fig. 10a).

The numerous branchings of its hyphae makes possible the enlargement of the fruit body on all sides (Fig. 10b); the browning of the border hyphae is limited only to the lower most base (Fig. 8) so that in general the pure white coloring is not thereby changed. The formation of the fruit bodies from their first origin up to the separation of the annulus from the margin of the cap lasts at least 14 days to three weeks, therefore a relatively long time in comparison to the development of many other large fruit bodies of the genera *Agaricus, Polyporus,* and so forth. The essential points differ completely in many respects from the description of the development of veiled fruit bodies of the genus *Agaricus,* as de Bary gives, so that a special description of this is in order. Hoffmann* gives a description of *Agaricus melleus* in which he limits himself only to the fully mature or almost mature condition. It is especially remarkable that he has not noted the relationship with *Rhizomorpha.*

De Bary says on page 68, forward, "The fruit body of several veiled *Agaricaceae (Agaricus campestris, A. praecox, Coprinus micaceus* and relatives) in its first stage is a body which is thickly and homogeneously interwoven from thin hyphae. Very early through differentiation the original homogeneous tissue of the main part of the fruit body is delimited and established. In the interior of the upper part of the body by a drawing apart of the tissue elements there arises a small and narrow air containing hole in the form of a horizontal ring. What lies above this goes to the cap, that tissue surrounded by and occurring below it goes to the stipe. That tissue on its outermost side corresponds to the cap margin; its hyphae carry on without interruption and without change into that of the stipe outer surface."

Investigation of the earliest condition of the fruit body of *Agaricus melleus* shows that also here, as in the case of the unveiled Agaricaceae, the cap arises through a flat annular groove (Ringfurche) which initially is fully open towards the outside (Fig. 17a), only later, by a fusion of the border hyphae of the cap and the stipe above the annular fold, is covered by a layer of hyphae, the veil. Even before the developing cap is detectable externally by a constriction below the apex (Fig. 6b and 8), the annular groove arises by the border hyphae of the fruit body being delayed in their development in a ring-shaped zone, while above and below it the border hyphae externally become very enlarged and lengthened and form characteristic end hyphae as are shown in Fig. 10a. The hyphae above the annular groove arrange themselves obliquely toward the bottom, those under the annular groove toward the top, so that in the beginning the formation of the annular groove is not noticeable externally. In Fig. 16, an enlargement of Fig. 15, the constriction already is detectable externally. The section (a) is shown enlarged (Fig. 17). While the border hyphae above and below the annular groove form a loose layer on the entire surface of the fruit body, the hyphae which push out to the annular groove are extremely fine (Fig. 17d), closely bound and pass in a nearly perpendicular direction towards the upper surface of the void. They form the first rudiments of the future hymenial layer. Because those hyphae which very early crossed each other in opposite directions in their growth orientation intertwine with one another and grow over the annular furrow (Fig. 20), this becomes subsequently closed towards the outside and gives the appearance that it arises in the interior of the fruit body. If one compares Fig. 20 with the illustration of a young *Agaricus campestris* given by de Bary *a.a.O.* Fig. 26, it appears from the agreement of the two figures to justify the supposition that also in the case of the latter fruit body in the region (r) only a subsequent fusion of the hyphae of the cap and the stipe has occurred.

---

*Hoffman, Icones analyticae.

[25]

The next stage of development is illustrated in Fig. 5, 18, 19, 20. The stipes are very thick, bottle-shaped, 2-5cm long, and bear a characteristically still very small head, whose surface is covered with light copper-colored and dark brown tufts of hair, while the former is white and appears striped, especially in the upper part, by dark tufts of hair. Section (a) of Fig. 19 is enlarged in Fig. 20. Above the annular fold (Ringfalte) the tissue of the fruit body has spread out horizontally into the cap substance (f). The margin of the cap has rolled itself toward the interior above the hole (g) which is enlarged also. In zone (i) the cap and stipe hyphae are growing through one another while in the angle, which the cap and stipe form, a layer of parallel running hyphae (h) branch towards the outside and by lengthening form the future annulus. The entire underside of the cap is covered by the hyphal layer formed from hyphae running at right angles across the surface. From the still smooth underside of the cap in Fig. 17 the lamella (d) have formed by a lengthening of the hymenial hyphae, resulting in radial stripes, which, however, do not pass completely from the margin to the stipe; as Fig. 21 and 28 show, in part go only from the margin towards the middle or even arise only on the outermost margin of the cap. Further development to the condition illustrated in Fig. 27 and 28 needs no further explanation. If one investigates the growing point denoted in Fig. 20i at the stage when the annulus is separating itself by tearing from the cap margin, one finds only a very small number of club-shaped basidia already equipped with spores (Fig. 22). If one isolates several hyphal branches of the trama with the basidia sitting thereon (Fig. 23), beside a club-shaped basidium (a,b) as a rule sits a small cylindrical cell (Fig. 23c) which is interpreted, indeed, only as a rudimentary basidium. Sometimes these lengthen beyond the common hymenial layer (Fig. 22c) and were earlier seen by Hoffmann but denoted incorrectly as (Pollinarien). To the supposition that these cylindrical cells are only vestigial basidia speaks the fact that these, just like the basidia, within 24 hours develop a germ tube if one permits a section through the lamella to lie in water under the cover glass on the microscope slide. Fig. 24 shows such a section through a lamella on which the club-shaped basidia frequently have developed even two or three germ tubes. The latter show septations and clamp connections just like the isolated hyphae of the rhizomorphs.

The spores (Fig. 25) are egg-shaped, show granular plasma or oil drops, and germinate very abundantly in water on the microscope slide within 20 hours (Fig. 26). Mostly they develop only one, sometimes also two germ tubes, which arise at any spot, but most frequently, indeed, in the vicinity of the pointed end. Spores sown out on 27 October already on the following day have developed germ tubes of the length indicated. On the following days numerous fungus threads, which appear to be arising from those germ tubes, developed on the microscope slide and abstrict small oval spores. Since I could not be satisfied about the relationship of these with the *Agaricus* spores, thus I have not made a drawing of this fungus. Unfortunately, I was not able to bring spores to germinate a second time, although to be sure the spores were taken from fully developed fruit bodies.

[26]

I also sowed spores on fallen beech leaves, which were placed in a flower pot containing moisture at ambient air temperature, I investigated spores which formed a white powder on the leaves and on the needles which had fallen in the forest, without being able to observe again a germination. It appears, therefore, that germinability is of very short duration and occurs only directly after the separation of the annulus from the margin of the cap.

The fully developed fruit bodies (Fig. 29) reach in part a considerable size. The stipe lengthens considerably, the cap reaches a diameter up to 15 centimeters.

In conclusion, I give Fries' diagnosis:

Pileo carnoso explanato squamoso-piloso, margine tenui expanso striato, stipite spongioso-farcto dein cavo elastico fibrilloso prope apicem annulo floccoso patente cincto, lamellis adnatis dente decurrentibus subdistantibus pallidis, dein albidofarinosos subrufescenti-maculatis.

Vulgatissimus inter folia decidua sylvarum ad basin truncorum, in pratis; etiam in tectis stramineis carbonariis, etc.

Color pilei melleus, senilis fuligeneus 1. olivaceus.

The pleasant odor, similar to *Agaricus campestris,* which is also characteristic of the mycelial body of this fungus, appears therefore to say that Fries, in contrast to the German authors who denote the "Hallimasch" as an admirable species generally used for food, is incorrect when he asserts that this fungus is poisonous, tough, and of unpleasant odor. It was too far beyond the problem which I have set myself to also investigate this question.

THE MODE OF LIFE OF AGARICUS MELLEUS

The fungus, whose mycelium and fruit body we become acquainted with, must be the most widespread and most frequent among the larger species.

In fall and early winter one sees the honey colored fruit bodies occurring usually in groups coming forth in hardwood and coniferous forests on the root stock of dead trees and on the stumps of felled trees. From the soil these appear here and there and by careful excavation their connection with a rhizomorph strand which is connected to a root or to a neighboring root stock can always be proven.

The rhizomorphs also appear very frequently outside of the forest in old moist woodwork, also, as already was mentioned at the beginning of this section, carried away with wood from the forest.

The fungus is able to nourish itself from already dead wood; it is under these circumstances a saprophyte; only for the earlier mentioned coniferous woods is it a true parasite.

As a saprophyte the mycelium brings about the destruction of wood when the rhizomorphs send numerous hyphae through the medullary rays into the interior of the wood, which hyphae spread into the entire xylem from which they nourish themselves. Here and there the hyphae come forth tuft-like from the wood and then form tufts of hair, entirely like those which appear externally on the rhizomorphs before the formation of side branches or fruit bodies. Since in such dead wood many other fungi likewise grow, thus often the fruit bodies of the latter have been ascribed to the rhizomorphs.

I found the fungus especially frequently on the stocks and root crowns of red beech, hornbeam, oak, birch, and European mountain ash.

In the case of all of these species the conditions indicate that the fungus is of a completely secondary character.

I have become doubtful only about cherry and plum, since I have observed often in plantations and along roads that a dead tree, which shows on the root stock abundant rhizomorph development and in the fall abundant fructification of *Agaricus melleus,* results in the death of neighboring trees in the course of the next year, which then likewise showed rhizomorphs on the root stocks as illustrated in Plate I, Fig. 26. In a row of young, vigorous cherry and plum trees, one tree after another withered annually without exception.

In the fall of 1872 a fruit body of *Agaricus melleus* which was arising from the soil mid-way between a tree which had been killed that year and a still-healthy neighboring tree standing about 3 meters away, proved that *R. subterranea* in its wandering at that time had come rather close to the still apparently completely healthy tree and in fact in the middle of the summer of 1873 the next tree died.

If also at present the observations support that among the hardwoods only species of *Prunus* can be killed by the fungus while in other hardwoods the fungus takes on only the role of a saprophyte, thus I still, however, have had too little opportunity to investigate thoroughly the effect of the fungus on hardwoods and I limit myself therefore to describing the disease of resin flux which occurs on the previously mentioned coniferous species.

This occurs from perhaps 5 years of age onward, in stands of Scots pine kills old stems still fit for cutting, in *P. strobus* and spruce up to 40 years, in fir, larch, *P. nigra, P. mugo,* and *P. pinaster* up to 20 years of age.

I have lacked the opportunity until now to confirm whether the disease occurs also in all conifer species, as is the case with Scots pine, at greater ages than those mentioned.

The disease begins while the plant in all parts appears completely normal and healthy, with the death usually of one side root.

The point from which the killing of this proceeds lies either near the end or near the root stock.

Very often I found such locally infected plants which still showed not the slightest disease in their foliage and which only had been rooted out and investigated since they stood in the proximity of a tree killed the previous year. In group-planted spruce in which one or more plants already have incurred the disease, one will find one or more of the still living plants in the condition that one or more roots have already been killed while the remaining are still completely healthy. Usually here the disease has taken as its starting point the spot which is in direct connection with an already dead plant, also as a rule the root collar. From here the disease can spread outward very rapidly to all roots, while in the case of individual plants the disease as a rule does not begin on the root stock but at some distance from the root stock on a side root and does not take on a severe character until the attacked side root is killed up to the root stock and from the latter the remaining roots also become attacked.

The killing of the roots is brought about by *Rhizomorpha fragilis* which bores into the roots (Plate I, Fig. 2), spreads out in all directions in healthy, living phloem as *R. subcorticalis* and thus from the point of attack continually approaches the root stock until this is reached.

On the root stock the rhizomorph spreads out in all directions until it has completely encircled it and now also can pass to the remaining roots of the tree. Plate II, Fig. 2 shows the debarked root stock of a still-living diseased 100-year-old Scots pine with a green-needled crown about 8 days after felling. The side root (a) had been attacked by *Rhizomorpha* for a long time, since the exposure of the root for a length of 3 meters showed that its wood from there towards the apex had already been completely rotted. The fungus had appeared growing in the living phloem at (dd) on the cut surface of a stump, while at (b), where felling had separated the bark slightly from the wood and as a result the phloem appeared browned, development was retarded. Here from the margin of the luxuriously growing *R. subcorticalis* arise only individual thin strands (Plate I, Fig. 4a), which grow upwards between the killed phloem and xylem, their fine ends also visible on the stump margin. The healthy condition of the phloem has resulted in a more vigorous development of the fungus than occurs in the recently killed stem part. All remaining roots and the side (c) of the tree were still completely healthy. The rhizomorph end growing in the living phloem causes a browning of those tissues with which it comes into contact.

First of all, the hyphae which lie in the jelly layer of the rhizomorph end must bring about the death of the cells, since death already appears where the threads, spreading out at right angles from the fungus body, still do not protrude beyond the jelly layer. The threads, then, by their development cause the formation of the resin holes above the root stock and the secretion of turpentine on the root stock.

From the root stock *Rhizomorpha* acquires direct access to all roots of the tree, in which it now develops likewise. It is difficult to determine how fast the growth of the fungus mycelium occurs.

According to Schmits* the freely growing rhizomorph end should lengthen daily about 2 mm. External conditions, especially temperature, exert without doubt a significant influence on this. In the case of younger, about 10-year-old plants, the entire disease does not last longer than about one year.

About 6-year-old Scots pines, artificially infected on the root stock in the middle of July, 1872, died in May, 1873 with all symptoms of the disease.

From the root stock outwards, the mycelium spreads not only to the still healthy roots but also upwards on the stem and to be sure the higher the thicker the tree. In most cases the further development of *Rhizomorpha* is limited by the death and drying out of the tree. In young plants we see the

---

*Ueber den Bau, das Wachsthum und einige besondere Lebenserscheinungen der *Rhizomorpha fragilis*, Linnaea, 1843.

[29]

mycelium ascend only one decimenter high, in older Scots pines ascending often about 2 meters high.

The death of all of the roots has as a self-evident result the drying out of the plant. During the disease, however, still other phenomena appear in the plant which can be traced back to the development of the simple threaded mycelium in the xylem. Irrespective of a most clearly detectable malady of the plant, explainable by the partial destruction of the roots, consisting of a yellow coloration of the needles, shortness of the shoots, and so forth, during the course of the year of the disease and root stock and the roots as far as these have already been attacked by *Rhizomorpha* show a heavy secretion of turpentine, which by oxidation is transformed to resin and which cements the nearest layers of earth in the vicinity of the roots (Plate II, Fig. 1).

As far as the phloem and cambium have already been killed by *Rhizomorpha,* obviously no formation of the annual ring occurs (Plate I, Fig. 21). Above the root stock, on the contrary, up to a height which is different according to the age of the plant, which in the case of 6-year-old Scots pine of about 2 meters height does not extend above 0.5 meters, one sees in the very small annual ring of the year of the disease in most cases a large number of completely abnormally formed resin canals (Plate I, Fig. 22b). Also in the bark often sizable holes filled with turpentine are detectable which in the case of spruce appear externally as boils (Fig. 22h).

In all remaining parts of the stem and in the twigs the annual ring is very thin to be sure but without any striking increase of the resin canals.

In the above mentioned artificially inoculated Scots pine the last annual ring was completely normal, both in regard to width and the number of resin canals, while on the root stock the characteristic resin secretion occurred as far as the mycelium had extended upwards in the phloem.

I have observed completely similar phenomena in many cases also in Scots pine and spruce which were killed by *Rhizomorpha* without artificial infection.

These cases are important because they give proof that an overproduction of turpentine in the earlier formed organs cannot be the cause of the disease and of the resin secretion. These are explained simply by the fact that when the plant has been attacked by the parasite at a time in which the annual ring essentially is already complete, as with Scots pine which was infected in July, no resin canals can arise anymore; this happens only if death did not occur before the next year's annual ring formation began.

Resin secretion and over-abundant formation of resin canals is explained, as was already mentioned earlier, by the development of the thread-like mycelium of *Agaricus melleus* in the xylem of a diseased plant. As Plate I, Fig. 14, 15, and 16 show, the fungus hyphae which arise from the marginal hyphae of the rhizomorph penetrate through the medullary rays into the interior of the xylem and pass in this way into the resin canals of the plant.

In these they grow upward and to be sure more rapidly than *Rhizomorpha* grows upward externally in the phloem, so that in the case of a plant which is diseased only on the lowest part of the root stock or only on a side root, investigation of the resin canals of the lower part of the stem can always already reveal the fungus hyphae.

The effect of these hyphae on the thin-walled and in most cases starch containing organs which surround the canals (Plate I, Fig. 23) is, however, one so intensive that soon a browning and finally a complete destruction of these cells occurs, as is shown (Fig. 25). Since also the medullary ray cells with their starch contents are completely destroyed, as Fig. 25 shows, thus the turpentine which is contained in the resin canals must flow outside. I believe also not to err when I assume that the product of the tissue and the cell contents destroyed by the fungus is essentially turpentine, since the significant amounts of secreted turpentine can scarcely be explained by the supply of turpentine already present in the interior before the disease.

The turpentine which is secreted on the side from the destroyed resin canals, partly through the likewise destroyed medullary rays, partly through the resin canals running in the latter where the bark on the roots and root stock is already killed by *Rhizomorpha,* is discharged freely between the wood and the bark and here fills up the cavities which arise after the death of the mycelium or by the shrinking of the bark, or it flows out freely into the soil when cracks arise in the bark by death and drying of the bark.

Above the root stock, however, where the phloem and cambium are still healthy, the turpentine flows laterally outwards from the inner destroyed canals of the wood through the medullary ray resin canals partially into the bark, and thereby produces the large turpentine boils (Plate I, Fig. 22h), and partly streams towards the cambium when this is growing. The significant influx of turpentine in the cambium produces numerous completely abnormally formed turpentine canals, consisting of large canal groups surrounded by irregularly thin-walled cells, as are shown in Plate I, Fig. 24. These also very soon after their formation are destroyed by the mycelial hyphae which come forth from the interior, so that they are very visible not only by their size but also by the brown color of the surrounding tissue. As Theodor Hartig has already observed, the rows of resin canals are sometimes fused together and form common cavities up to the size of a bean which then are thickly filled with pitch.

Already from this circumstance the conclusion can be drawn that a discharge of turpentine from the interior to the cambium is the cause of the abundantly formed resin canals.

Now that I have proven how the resin secretion on the lowest part of the stem and on the roots, and how the abundant occurrence of turpentine canals in the bark and in the annual ring in the year of the disease can be attributed to the development of the mycelium in the interior of the tree, a few remarks about the progress of the disease are still necessary.

If the disease begins on or near the root stock, so that in a short time the mycelium is able to spread itself over all roots of the tree, then the disease progresses very rapidly. Without the occurrence of previous stunting, shortening of the shoots, or bleaching of the needles, the trees die suddenly by drying, after the destruction of the larger part of the roots has essentially interrupted nutrient uptake. Often on plants which are in their most vigorous growth the young shoots suddenly hang down limp and are rapidly dried out.

In other cases, the disease takes a slower course and to be sure when a side root somewhat removed from the root stock is diseased. A longer period of time passes before the mycelium reaches the root stock and before the remaining roots are killed to bring on the drying out of the plant. During this period of the disease, as a result of the gradually increasing root destruction, there occurs a bleaching of the needles; the decrease of nutrient conduction has the result that the new leader and the annual ring form only poorly. Very often one finds plants which by their green although stunted foliage can be recognized as living, whose root stocks are entirely covered with resin, and on which most or all of the roots have been already killed. In spring and summer after the death of the roots, the drying occurs very rapidly, especially during bud break; in the fall with continuous moist weather young trees can remain rather fresh and green for weeks after the roots have been completely killed.

The cylindrical, dark mycelial cords of *R. subterranea* occur always more or less numerously, clinging to the killed roots or surrounded by the cemented together mass of earth. Their importance to the disease will be discussed later.

On trees killed during the summer, in early October begins the development of the fruit bodies which has been described in detail before. Towards the end of October occurs the unfolding of the cap. At the end of November many fruit bodies still occur although they are already dark colored. Remains of the fruit bodies often can still be seen in the following spring on the root stocks of dead trees.

On older Scots pine with thicker bark and on eastern white pine, I found the fruit bodies occur only between the wood and bark on cut stumps (Plate II, Fig. 3). The bark here cracks with too much difficulty to allow the fruit bodies to arise on standing trunks. The fruit bodies arise here mostly on the points of cord-shaped rhizomorphs and appear individually or in groups at some distance from the tree sprouting upward from the soil. In spots where one or more years before an old Scots pine killed by the fungus has been felled, thousands of fruit bodies appear in the fall in part sitting directly on the shallowly running roots and in part springing from the round rhizomorph cords. The mycelium nourishes itself for a series of years as a saprophyte on the killed plant, so that one encounters the fruit bodies in spots in which the disease already has attacked around itself for a number of years, also on small stems already dead for five or more years.

In few plant diseases does the infectious character appear so without doubt as in that caused by *Agaricus melleus*.

In thick stands of plants, such as we encounter as a rule in young coniferous stands, in the neighborhood of a plant killed by the fungus, in the course of years more plants always die, so that in the most vigorous fully-stocked plantation gradually large openings or clusters of dead plants form, in which those plants standing next to the periphery have been killed last, those plants standing in the middle have been killed first.

This observation has permitted the practitioner to give the disease the name

[32]

"Erdkrebs" (ground canker) in order therein to express the cancerous spread of the disease. With the row-wise placing of the plants, the killing occurs from the beginning point outwards on both sides, so that with considerable certainty the death of the plant which will follow shortly can be determined before the slightest symptom of the disease can be detected. In less thick stands of plants in the course of years the spread of the disease can be detected also, yet many plants remain spared and there occurs more of a thinning of the stand.

There can be, therefore, indeed, no plant poison spreading itself outwards in the soil gradually from one point; that surely would wipe out the plants uniformly.

Further, very noteworthy is the fact that a tree killed by the fungus infects just as easily trees of other coniferous species in the vicinity as it does individuals of the same species.

In mixed forests eastern white pine becomes diseased in the vicintiy of killed spruce, larch and Scots pine; spruce in the vicinity of killed Scots pine; larch are infected by Scots pine or spruce, and so forth. This transmission of the characteristic disease from one species of tree to another can be explained neither by climatic influences nor yet by the soil; on the contrary, it is completely explainable by the fact that the true parasite which causes the disease makes no distinction between the above named conifer species.

A conclusive fact about the character of the fungus, if the portrayed life cycle still permits doubt to occur at all, must be the fact that of four about eight-year-old Scots pines which I infected on the 13th of July, 1872 by placing bark pieces with *Rhizomorpha subcorticalis* on the root stock freed from the earth and on a freely situated but completely uninjured side root, in May, 1873, before the breaking of the buds, two died and abundant mycelial development appeared particularly in the proximity of the infected spots. On both plants remaining healthy, the mycelium utilized for the infection was completely dead.

It was probable that the mycelium for reasons unknown to me at the present time rotted very soon after its utilization so that its effectiveness soon decreased to nothing.

This permits scarcely a doubt that infection in nature is brought about by the round cords of *R. subterranea* spilling out from a dead plant into the soil in all directions. One finds these at a depth of up to one decimeter growing out in a horizontal direction and with great care can expose these for long distances. Whether the penetration of the rhizomorph into the healthy roots of neighboring plants is made possible only by certain conditions, by small wounds, and the like, or whether the rhizomorph point is able to penetrate at any point on the healthy root I must leave still undecided for the present. The fact is sufficient that in thick stands of plants none escape infection and death.

After what I have said previously about the disease it now must not be expected of me to refute all those objections which have been raised and certainly will be raised purely from prejudice against the expressed opinions. However, in order to cut off from the start a whole series of possible objections, should here only be emphasized that the climatic conditions have no visible influence on the occurrence of the disease, that the dying occurs in spring, summer, and fall, perhaps also in winter, that the disease occurs on the best and the worst soils, on wet and dry soils, on heavy and light soils.

Only one observation will I emphasize; namely, the fact confirmed repeatedly by me that the disease often occurs especially destructively where the planting of conifers has been carried out after the felling of hardwoods or where numerous hardwoods have intermixed between the conifers. This phenomenon is probably due to the fact that on the roots of the hardwood stumps remaining behind in the soil rhizomorphs have abundant opportunity to develop and spread from these over to the conifers. But it should not be maintained from this that the rhizomorphs attack only from hardwood stumps to the conifer woods since, as we said earlier, the mycelium grows for several years on all conifer stumps and roots; therefore, hardwood stumps are not necessary for the spread or origin of the disease.

The condition of the plant itself likewise gives no basis for the supposition of other causes of the disease than the penetration of the parasite. Without distinction very well-growing or very slow-growing trees and dominant and suppressed plants die of the disease. The preceding year is of no influence at all; the disease occurs in one year rather in the same manner as in other years.

I do not think that it is a question of a disease which has resulted only in modern times or which is occurring more severely than earlier; on the contrary, it is generally distributed but was not recognized or was attributed to other influences. Bad soils, influences of the weather, and above all insects were regarded as causes of the dying. That larvae of *Melolontha melolontha* had eaten off the roots, *Curculio notatus* had colonized the plant, the effects of *Hylesinus piniperda* and the likes soon observed as a rule in the bark after the sickening of older Scots pine, were considered as causes of the disease, while they are only a symptom of the disease which is occurring. In many cases, however, one also recognized that it was a question of a special disease which one could explain but without publishing about it in the literature. I myself have observed the disease for 15 years in wide distribution in all Germany.

As a voucher for the fact that the disease was known in practice even before the first publication, I publish a brief notice from my daily logbook from 1859: "In the district Oberkochen (in the Swabian Alps) occurs an unexplainable disease of spruce in great extent. Both in pure spruce plantations and in beech regeneration which has been replanted with spruce continually a great amount of spruce die individually and in groups."

"In the first year they become yellow, in the second they lose their needles and become dry. As a result of this the most unbroken spruce plantations gradually become porous with clearings. If one investigates the plants, one finds not the slightest wounding on the root crown or on the roots. Only the larger number of resin boils is noticeable which occur on the lower part of the stem as well as especially on the root crown, by which the latter sometimes is quite swollen. Any basis for the disease in the soil and so forth is not known."

According to a notice sent to me by the late Geheimrath Ratzeburg, already in 1847 in Neunkirchen forest district in the Trier administration district the disease had been known in the pine stands for a long time and "mixed spruce was affected and killed by this disease."

From my audience I obtain numerous reports about the occurrence of the disease. In 1868 the present forest supervisor candidate, Fr. Boden, sent me from Mollenfelde near Göttingen samples of *Pinus sylvestris, P. strobus, P. pinaster, Abies excelsa, A. pectinata,* and *Larix europaea* all of which had been killed by the disease, with the report that, "the fungus occurs in equal amounts on all sites, on slopes and bluffs, as well as on the deep-soiled, extremely fertile portions of red marl, variegated sandstone and shell limestone. Likewise, the culture of the stand, either through seed or by planting and the age up to 20 years seems to be without significance."

Forest supervisor candidate Schwieger from Morbach informed me that the disease in question is especially widespread in the forest district there in spruce plantations, especially of 5 to 6 years of age.

Finally, I report what forester Schimmelpfennig at Osnabrück wrote in May, 1870 about the occurrence of the disease in a ten-year-old *Pinus strobus* stand at the forestry district of Iburg in the forestry compartment Thorensundern:

"The surface had been heavily sodded up to 40 years ago, as unfortunately many of our local forest sites, then sown in rows with spruce, which because of permanent stunting were felled in 1859-60 at an age of 25-30 years. From 1860 and later followed a planting with eastern white pine, partly pure, partly mixed with spruce. The eastern white pine have been diseased for seven years and small bark beetles occurred in the stem; thus the latter until now were known as the plantation destroyer, until I thoroughly examined the stand and soon discovered the main trouble. Twenty percent of the stand has already been killed and I have little hope that the rest is to be preserved, although the diseased stems are immediately taken out and burned, etc."

The previous reports are sufficient to afford proof that in practice the disease is very well known and had been known for a long time before reports about it were published in the literature.

Finally, it must still be mentioned that as countermeasures against the spread of the disease, pulling up or digging out of the dead plants is to be recommended.

It is frequent in practice, when the infectious character of the disease was known and the cutting of the dead plant was carried out with care, but self-evidentally without any kind of result, because one has left in the soil the tree part in which the parasite occurs and nourishes itself. Because the rhizomorphs develop for several years on the root stock of the dead plants, and the cords spread further from these outwards, an immediate pulling up or especially a digging out of the dead, small stems not only removes the source of nutrition of the parasite but also removes the greatest part of the parasite itself. Annihilation of the rhizomorphs is difficult and almost impractical only where numerous killed roots and root stocks of hardwoods afford the rhizomorphs a great opportunity for development once these are already present.

The opinion of one practitioner* that the fungus theory is so infinitely wretched, because indeed we cannot act against the fungus, is incorrect, because with complete knowledge of the causes of diseases themselves in most cases countermeasures can also be found. If, however, such are not found, thus there still lies therein no justification to maintain as incorrect the explanation about the effect of the fungus or to shut ones self off completely from the results of scientific investigations.**

---

*Reuss. Lärchenkrankheit.

**Some further reports about *Agaricus melleus* follow in the appendix.

Explanation of the Illustrations

Plate I. The Mycelium of *Agaricus (Armillaria) melleus*

Fig. 1. Abundantly branched cords of *Rhizomorpha fragilis* between the bark plates of an old Scots pine.

Fig. 2. Root of a Scots pine killed by *Agaricus melleus*. A cord of *Rhizomorpha fragilis* var. *subterranea* circles the root with its branches. At (a) several branches bore into the bark, in which they develop further as *R. fragilis* var. *subcorticalis*.

Fig. 3. A somewhat flattened cord of *R. fragilis subterranea* which was growing upward between the xylem and phloem of an old Scots pine after its death. Numerous branches extending at nearly right-angles on the right side up to its rounded off end retain a dark brown rind and uniform thickness, while on the left side the side branches run out into extremely fine points, almost rindless, and therefore colored white and shows that structure on the end which is enlarged in Fig. 6-9.

Fig. 4. The end of a *R. fragilis subcorticalis* growing in the healthy phloem of a living old Scots pine (part of the sample is shown reduced in Plate II, Fig. 2). Single, round fungus cords (a) arise from the margin of the fungus body which developed in a broad, ribbon form at (b).

Fig. 5. Flat-spreading *R. fragilis subcorticalis* developed in the phloem of a living young Scots pine. The margin is abundantly indented (b) by single round cords (a); this form stands in undoubted connection with *R. fragilis subterranea* which develops in the soil. At (c) root-like branched cords arise from the margin, whose outermost ends have the form illustrated in Fig. 6-9.

Fig. 6. The end of an extremely fine rhizomorph cord (Fig. 5c) consisting of several parallel-running hyphae.

Fig. 7. Point of branching of a thread-like rhizomorph cord. The hyphae with clamp connections.

Fig. 8. Transition of the cord-shaped mycelium of *Agaricus melleus* to simple threaded mycelium. The hyphae often with a granular surface.

Fig. 9. Slightly enlarged end of a rhizomorph branch of Fig. 5c; the hyphae are held together by a colorless jelly.

Fig. 10. Young end of *Rhizomorpha fragilis*. The parallel-running hyphae in the growing point (a) are extremely thin, small, and uniform. Behind the end the cells of the inner hyphae (b) enlarge while the external hyphae (c) increase only slightly in size. The end is surrounded by a gelatinous coating (d) in which numerously branched hyphae lie embedded. These arise on the outermost point as branchings of the parallel boundary hyphae. Behind the point rise somewhat thinner hyphae (e) extending at right angles from the cord, which grow out beyond the boundary of the jelly layer and are numerously branched.

Fig. 11. A part of Fig. 10x enlarged. The hyphae of the future inner rind (c) (Fig. 19b) which remain thin walled are surrounded by the hyphae of the outer rind (d) (Fig. 19a) which later show very thick walls. In the jelly layer lie irregularly running many-branched hyphae (e), while from the marginal hyphae (d) arise hyphal cells (h) which extend out at right angles.

Fig. 12. Jelly with hyphae taken from behind the growing point (Fig. 10d). Sometimes there occur therein characteristic vesicular swellings of different forms.

Fig. 13. The hyphae originally embedded in jelly (Fig. 11e) frequently one cm behind the point (Fig. 4x) show altered cell walls. By continued stretching of the hypha after the ceasation of cell wall formation there occurs a tearing apart of the latter, whose broken pieces are held together by the plasmalemma (Innenschlauch).

Fig. 14. A radial section of the wood of a diseased but still living Scots pine, taken from a spot in which *Rhizomorpha fragilis subcorticalis* has invaded the living phloem for only a few days. At left a part of the rhizomorph, at right the xylem. The central hyphae of the fungus body (Fig. 10b) have enlarged significantly. The individual hyphal cells (Fig. 14b) have put forth tender, only sparsely septate, numerously branched hyphae (g) which are growing parallel to the longitudinal axis of the cord, pulling the cells apart and finally form the so-called core of the rhizomorph, in which the mother cells of the hyphae are to be found only with difficulty (Fig. 17b). The cells of the inner rind (c) send forth no hyphae, the hyphae of the outer rind (f) send forth hyphae which extend at right angles (h), and usually penetrate only through those starch-bearing medullary ray cells of the wood provided with a plasmalemma. Only seldom do individual hyphae grow into the wood fibers. The jelly and its hyphae have killed the phloem and cambium at their first contact, which shrinks as a result of this, and the intermingling of the hyphal jelly with this forms a brown, shapeless layer (i).

Fig. 15. Radial section through a Scots pine root killed by *Rhizomorpha subcorticalis*. The killed, browned bark (b) is separated from the interior part of the phloem (i) and xylem by the wavy bent fungus body. The rhizomorph shows the white core (g) and the brown rind (cc). From the outermost hyphae of the latter arise numerous hyphae (hh) which, growing throughout the shrivelled up phloem and cambial area, form between this and the xylem a thin zone (k) of irregular, entangled hyphae which penetrate almost exclusively through the medullary rays into the interior of the wood.

Fig. 16. Cross-section through the same object more greatly magnified. The core hyphae of the rhizomorph (gg) are cut through at right angles, the inner rind (c) forms a thin-walled pseudoparenchyma (Fig. 17c). The outer rind (ff) consists of non-browned, thinner hyphae, surrounded by a jelly layer, whose right-angled branching hyphae (hh) individually or in groups likewise are surrounded by the jelly. The shrivelled up phloem and cambial area (i.i.) are penetrated by these. At (k) these to a certain degree seek the openings of the medullary rays in order to penetrate into them.

Fig. 17. Longitudinal section through the rind and a part of the core of *Rhizomorpha fragilis subterranea*. The pseudoparenchyma of the rind (c) shows toward the outside an outer rind (f) of narrow and thick-walled rows of cells. The jelly layer and the hyphae extending at right angles are absent. The core consists of the sparsely septate and branched thin hyphae (g), which arise from

[37]

the cells (b). Where a side branch or fruit body arises on a rhizomorph, from the innermost rind cells arise numerous hyphae which are distinguished from the core hyphae by the larger thickness, frequent septations and branching, as well as by the clearly double-layered cell wall. They form a pseudoparenchyma (mm) which breaks through the rind and grows forth as a young fruit body or side branch (Plate II, Fig. 7, 14m).

Fig. 18. Cross-section through the rind of *Rhizomorpha fragilis subterranea*. The outer-rind (a) consists of narrow-lumened hyphae with thick walls; the inner rind (b) of wide-lumened thin-walled hyphae. At (c) the core area begins.

Fig. 19. Cross-section through a somewhat thicker rind of a rhizomorph cord after treatment with potassium hydroxide solution whereby the boundaries of the individual cells clearly appear.

Fig. 20. Cross-section through the root stock of a four-year-old Scots pine. The last annual ring is not fully formed but without numerous resin canals. Under the bark appear small, white ribbons of *Rhizomorpha subcorticalis*.

Fig. 21. Cross-section through the root stock of a young Scots pine killed by *Agaricus melleus*. The annual ring of the last year of the disease is entirely absent here, the ring of the previous year (a) is entirely normal. Between the wood and bark just as in bark cracks, abundant resin has been secreted (h), while at (rr) cut-through ribbons of the rhizomorphs can be seen.

Fig. 22. Cross-section through the same Scots pine, three decimeters above the soil. In the bark appear large resin cavities (h). Between the wood and the bark no rhizomorphs are to be seen. The annual ring formed in the year of the disease (bb) is very narrow and distinguished by numerous, very large resin cavities. In the resin canals especially of the inner annual rings, numerous threaded mycelium and little resin.

Fig. 23. Normal healthy resin canal of Scots pine. The resin duct (k) is filled with resin, surrounded by the secretory cells (a) which are surrounded by thin-walled, starch-bearing cells (b).

Fig. 24. Resin canal group from the annual ring of the year of the disease. The resinless cavities (kk) are surrounded by killed, collapsed cells. The mycelium of *Agaricus melleus* is seen in the cavities.

Fig. 25. Completely destroyed resin canal (k). The thin-walled tissue on its margin has disappeared. Some mycelial residues are still to be seen. The adjacent medullary ray is likewise completely destroyed.

Fig. 26. Cross-section through the lower stem of a killed cherry tree with *Rhizomorpha fragilis subcorticalis*. The xylem (a) is separated from the phloem (b) by a thick brownish jelly layer, in which in different directions ribbon-shaped strands of rhizomorphs (dd) are embedded. From the rind of the latter arise numerous hyphae which extend at right angles. The jelly is penetrated by branched hyphae in the same manner as in Fig. 11 and 12.

Plate II. The Fructification Organs of *Agaricus (Armillaria) melleus*.

Fig. 1. An eight-year-old Scots pine killed by *Agaricus melleus*. The mycelial strands of *Rhizomorpha fragilis subterranea,* which cling to the roots and arise from them, are partly sterile (a.a.), partly fructifying (b.c.). At (b) numerous fruit bodies have developed along the two forked strands, only a few of which, however, have reached complete development. At (c) a cord has branched many times and bears fruit bodies on the tips of two branches, while the remaining branches show only vestigial fruiting bodies. *Rhizomorpha fragilis subcorticalis* growing under the bark of the root stock has sent forth through bark cracks numerous fruiting bodies (d). The bark in the vicinity of the tap root and the root crown is cemented by secreted resin.

Fig. 2. The root stock of a living 100-year-old Scots pine a few days after felling illustrated without bark. The side root (a) shows abundant development of *Rhizomorpha subcorticalis,* which spread out fan-like on the root stock and grew in the living phloem (c), while on the contrary at (b), where the bark had been somewhat loosened by felling and had died as a result of this, was transformed to the round cord-shape. A part is illustrated in natural size (Plate I, Fig. 4). Where the mycelium growing in the healthy phloem has reached the cut surface of the stump, it is growing freely upward above this in the form of broad lobes (d.d.). The tap root and remaining side roots, as well as the side of the stem (e) are still completely healthy.

Fig. 3. Stump of a Scots pine killed in July, and felled at the beginning of August, in October of the same year with abundant fructifications of *Agaricus melleus.* The fruit bodies appear partly (a) between the bark and wood on the margin of the cut surface, partly (b) on the margin of wounded faces of the roots, partly (c) at greater distance from the stem on the ends of strands of *Rhizomorpha subterranea.*

Fig. 4. Young Scots pine on whose root stock young fruit bodies of *Agaricus melleus* are breaking forth (8 October). The fruit bodies appear partly in clumps on bark cracks on the margin of *Rhizomorpha subcorticalis* (a.a.), partly individually or in pairs on the ends of outward growing rhizomorph cords (b.b.).

Fig. 5. Further developmental stage of fruit bodies of *Agaricus melleus* on a young Scots pine, whose bark in part has been artificially removed in order to show *Rhizomorpha subcorticalis* (a) and its upper margin at (d). Individual cords of *Rhizomorpha subterranea* (b) arise from the margin of *Rhizomorpha subcorticalis.* The fruit bodies (c) show already clearly the still button-shaped and rust-colored caps.

Fig. 6. The youngest condition of the fruit bodies of *Rhizomorpha subterranea,* which are individual at (a) and (b), paired at (c) or formed in tufts.

Fig. 7. A young fruit body (Fig. 6a) enlarged and shown in cross-section. The rind of the rhizomorph cord (e) is colored brown towards the outside, the core (g) is white and consists of nearly parallel running hyphae (Plate I, Fig. 17). On the surface of the rind here and there tufts of hair (n) occur under which a pseudoparenchyma (m) forms (Plate I, Fig. 17m), which breaks through the rind and forms the tissue of the young fruit body (m.m.).

Fig. 8. Young fruit body (Fig. 6b) enlarged and shown in cross-section. The core (g) of the rhizomorph cord is dislodged by the pseudoparenchyma (m) which arises on the inner side of the rind. The marginal hyphae of the fruit body bow obliquely towards the outside. An annular fold (a), the first rudiment of the cap, forms on the tip.

Fig. 9. Hairs on the rind of the rhizomorph with a granular surface, sparse septations, and single cells with clamp connections.

Fig. 10. A part (x) of Fig. 7 enlarged. Single hairs (n) similar to those of Fig. 9 arise from the pseudoparenchyma, which sometimes as at (o) shows a characteristic ball-shaped cell on a short side branch. The tissue of the fruit body consisting of numerous here and there bent cylindrical hyphal cells (Plate I, Fig. 17m) in the outer region (m), where numerous branches bend towards the outside, shows variously shaped end-members (a.a.) while the still growing hyphae stretch and branch frequently (b.b.).

Fig. 11. A many-branched cord of *Rhizomorpha subterranea,* whose base transforms into a broad, ribbon-shaped *Rhizomorpha subcorticalis* (a) which has developed under the bark. The branches show numerous tufts of hair (c) and bear on the tips more or less developed fruit bodies (d.d.).

Fig. 12. A part of Fig. 1b at natural size. Towards the top the tufts of hair are gradually transformed into the fruit body.

Fig. 13. Cross-section through the cord shown in Fig. 12x.

Fig. 14. A part of Fig. 13 enlarged. The core (g) is surrounded by a thick rind colored dark brown towards the outside. On this tufts of hair (n.n.) arise externally; on the inner side under the tufts of hair a pseudoparenchyma (m.m.) develops (Plate I, Fig. 17m), which separates partially (x) from the pith tissue and at those places gives rise to the rind.

Fig. 15. Young fruit bodies coming forth between bark cracks from *Rhizomorpha subcorticalis*.

Fig. 16. The middle of these fruit bodies enlarged. The cap is already clearly recognizable by the fact that above the annular fold (a) the tissue has developed vigorously.

Fig. 17. A part of Fig. 16 enlarged. The annular fold (a) towards the outside is still not complete and is formed by very fine marginal hyphae bending bow-like towards the surface (d.d.). Above the annular fold the large margin hyphae of the cap (c.c.) grow obliquely towards the lower side, under it the marginal hyphae of the stem (b) grow obliquely towards the upper side, in order as a result to grow into one another and to shut off the hole externally. The tissue of the stipe is designated with an (e), that of the cap with an (f).

Fig. 18. Fruit body with a distinct, still undeveloped cap.

Fig. 19. The same in section. Above the annular fold at (a) a distinct velum has developed.

Fig. 20. The section (a) of Fig. 19 enlarged. The annular fold (a) is filled with air, in this lamella (x) are growing out from the upper surface. The hymenial layer (d) covers not only the lamella but also the entire underside of the cap. From the tissue of the cap (f) and the turned-over margin of the cap (g) arise large-celled hyphae (c) which, growing downwards, below the annular fold at (i) cross with the upward growing marginal hyphae (b.b.) of the stipe (e) and unite with them until, after complete formation of the cap, they are torn here from one another again. From the hyphae of the stipe below the annular fold numerous, completely parallel running hyphae (h) extend towards the outside and by their lengthening form the armilla.

Fig. 21. Horizontal section through the young cap (Fig. 19). The designation of the individual parts is the same as in Fig. 20. All the lamellae do not run from the margin of the cap (g) to the stipe (f), but in part run only from the margin of the cap up to the middle or occur only on the margin of the cap.

Fig. 22. Cross-section through a cut lamella. The hyphae of the lamella substance (d) branch frequently and send side branches towards the outside, which end in club-shaped basidia (a.a.). On a great part of these usually arise four sterigma, whose ends swell to spores (b.b.). Single thread-like vestigial basidia (c.c.) extend beyond the hymenial layer.

Fig. 23. Individual hyphae of a lamella with the basidia developed thereupon, which remain partly sterile (a), partly bear sterigma and spores (b), partly vestigial (c) (Hoffmann's Pollinarien).

Fig. 24. A section of a lamella kept a few days in moist air. All basidia including the so-called Pollinarien have developed one to three germ tubes.

Fig. 25. Mature spores.

Fig. 26. Germinated spores.

Fig. 27. Fruit body during the separation of the armilla (h) from the margin of the cap. At the base of the stipe still appear some vestigial fruit bodies, all of which sit on the end of a rhizomorph cord.

Fig. 28. The same fruit body in section. The armilla (h) already separated on the right side from the margin of the cap and fallen away; on the left side it still stretches tightly between the margin of the cap and the stipe. The lamella (d^x) run from the margin only a short distance, or up to perhaps the middle of the cap underside, or to the stipe.

Fig. 29. Completely formed fruit body.

Fig. 30. The base of the fruit body in Fig. 29 shown in section and enlarged. The rhizomorph cord as in Fig. 6c has developed numerous fruit bodies on the end, of which, however, only the middle one has completely formed, while the rest are vestigial. The pseudoparenchyma of the fruit body (m) towards the underside gradually transforms to the core (g) of the rhizomorph cord.

Fig. 31. Base of a fruit body, which has arisen from the fusion of several fruit bodies.

Trametes Pini Fr.

(Plate III, Fig. 1-19)

The Pine Fungus (Kiefernbaumschwamm)

Causal organism of red-rot, bark-, ring-, or heart-shake of Scots pine.

The symptoms of the disease and decay of wood which are known as red-white-, heart-, branch-, butt- or stem-rot, as bark- or ring-shake, and by still other names, notwithstanding their great importance for the forester and technician until now have not been thoroughly investigated. In practice one designates those symptoms of wood decay which are distinguished by their darker coloration as "red-rot", those decay conditions in which the white color predominates or shows here and there as "white-rot". According to the part of the stem in which the decay occurs, it is designated as heart-rot, branch-rot, butt-rot, stem-rot and so forth.

All of the designations are only collective names for in part very different types of diseases and Willkomm in his work, "Die Roth- und Weissfäule", has committed the great error, in that he extended his observations on spruce to almost all wood species and wood diseases, generalizing these without having obtained through investigations the conviction that he was justified to do this.

The red-rot, bark-, ring-, or heart-shake of Scots pine has not the slightest similarity with the red-rot of spruce.

Theodor Hartig is the only one who has scientifically investigated this disease until now, and to be sure in his, "Abhandlung über die Verwandlung der polycotyledonischen Pflanzenzelle in Pilz- und Schwamm-Gebilde und der daraus hervorgehenden sogenannten Fäulniss des Holzes." Berlin bei Lüdertiz 1833.

In consideration of the state of science and especially of mycology at that time, the achievement of Theodor Hartig is to be considered a high one, which he earned by the fact that he first proved that in this disease the fungus formation played a prominent role.

The opinion laid down in the designated treatise about the nature of the disease can be summarized briefly in that the decay of the wood is brought about by the age of the tree part and the functionlessness of the plant part which thereby occurs. This functionlessness can also occur earlier under some circumstances, such as a result of unsuitable site or weather conditions or other harmful influence.

Cell content or cell walls of the dead organs disintegrate into the smallest membrane globules (Monaden), which transform themselves directly into fungi.

Through an arranging of the Monads in a series with one another fungal threads arise which, by their further development, are able to make still healthy wood sick and to destroy it. To the fungi which arise through spontaneous generation in this manner in the interior of the xylem Theodor Hartig gave the name "Nachtfaser (Nyctomyces)", and designated these as an independent genus of fungi, which are not related to any spore bearing fruit body. Of greatest interest is the fact that Theodor Hartig on page 39 of the designated treatise had already stated that his opinion about the origin of the fungus by spontaneous generation would be entirely untenable if a connection of the mycelium in the interior of the tree with the fungus which comes forth from the branch holes could be proven.

It says word for word: "One of the most interesting observations was furnished me by a piece of an old Scots pine, in which a broken off, non-healed-over branch stub is growing 5 to 6 inches deep into the wood and several inches protrude out of the bark. The outer part of the branch has been completely denuded of bark. On the underside of this, with the hymenium turned toward the earth, sits a *Boletus* whose context is leathery and red-brown. The part of the branch which is sticking in the wood, mainly towards the heart of the tree, is surrounded by the Nachtfaser, to be sure in a state of perfect formation, as a brownish-yellow wooley mass.

Naturally this must lead me immediately to the thought that there occurs some connection between Nachtfaser formation and the external fungus, ...to the thought that the *Boletus* is the product of the outward growth of the Nachtfaser under the influence of the external atmosphere. If these views had been confirmed, so thereby would all the bases which had been established as to the proof of the spontaneity of each fungus form have become invalid. We could then consider the Nachtfaser not as an independent fungus but as a fungus which, developing in the wood and growing forth from this base, produces in light another higher developed fungus".

It is in fact very much to be regretted that at that time Theodor Hartig was not successful in convincing himself of the relationship of the mycelium with the fruit body.

In the knowledge of the nature of diseases of wood we would then probably be very significantly further ahead than we actually are today.

The work of Willkomm: "Zur Kenntniss der Roth- und Weisfäule" in his "Mikroskopische Feinde des Waldes" Dresden 1866 has the merit of having turned the attention of researchers again to diseases of wood.

In the first chapter: "Geschichtliche Darstellung und kritische Beleuchtung der bisherigen Ansichten, Meinungen und Hypothesen" (Historical presentation and critical examination of the opinions, thoughts, and hypotheses up to now) the author gives in 32 pages a summary of pertinent literature, which so much the more saves us the trouble of investigating these here once again, as from this the complete uselessness of the works on this subject up until now is apparent and these seem to justify the question whether Willkomm did not go too far in taking up the time of the reader by publication of all these opinions.

As for the work of Willkomm himself, I reserve my judgment until the time when I have studied more thoroughly the red rot of spruce than I until now have had the time and opportunity.

In any case, the investigations of Willkomm are very valuable and offer a rich abundance of interesting observations. These do not, however, develop to a satisfactory conclusion; on the contrary, in many directions they require completion, confirmation or correction. The origin of hyphae of *Nyctomyces candidus* from the spores of *Xenodochus ligniperda* arranged in a row with one another which Willkomm describes and illustrates, call to mind vividly the description which Theodor Hartig gives for the origin of Nachtfaser from Monads.

Since this observation stands in contradiction with all of the laws of growth, in any case it still requires confirmation.

The proof as to the manner in which the spores which exist in the sporangia of *Rhynchomyces violaceus* reach the interior of the wood fibers is entirely lacking and Willkomm plainly errs when he holds these as identical with those spores from which *Xenodochus ligniperda* arises.

Probably Willkomm was confused by the fact that besides spruce he also made several observations on other wood species which were infested by entirely different diseases.

Willkomm obviously did not know the disease of Scots pine worked upon by Theodor Hartig and known as bark-, ring-, or heart-shake, also as red-rot, which is so generally distributed and perhaps possesses the same importance for Scots pine stands which in spruce stands must be attributed to the red-rot caused by *Xenodochus*.

On the other hand I will here similarly establish that the red-rot in Scots pine described by Willkomm must be very rare since up to now I have never observed it in spite of careful search.

The red-rot of Scots pine is caused by the growth of mycelium of *Trametes Pini* in the heartwood, which, utilizing the unhealed over branch stubs as breaks through the sap wood, produces the bracket-shaped fruit bodies outside of the stem in the branch holes. I place the description of the parasite ahead of the description of the symptoms of the disease.

THE MYCELIUM

The mycelium of *Trametes Pini,* like that of *Agaricus melleus,* is distinguished by great diversity of form and color.

It occurs not only as a simple threaded brown-colored or colorless mycelium growing in the wood cells, but also in the form of large matted fungus bodies and fungus skins. The latter form primarily where cavities in the form of cracks or splits arise in the diseased wood, where, by destruction of the medullary rays, an abundant development of hyphae becomes possible. Colored predominantly rust brown, the mycelium in some circumstances even takes on a white color.

The simple threaded mycelium in a youthful condition, as can be observed in artificially infected trees (Plate III, Fig. 16) or on the border between healthy sapwood and the diseased heartwood on already long-diseased trees, shows septations only extremely rarely but shows abundantly branched hyphae filled by only slightly opaque and colorless plasma. As long as these colorless hyphae still bear sap, a double layering of the cell wall cannot be detected (16 a.b.); this appears very clearly as soon as the hyphae are empty (17b).

The hyphae have predominantly a thickness of 3 microns. From them arise, however, already in the first stage of the disease, single thin side branches (16c, 17e) whose diameter often scarcely reaches 1 micron. The thick and thin hyphae bore at random through the walls of the wood cell with ease, choosing the pits for passage only rarely. From the thicker hyphae often arise very numerous short branches which bore through the walls without developing further in the neighboring cells.

After attaining a certain age or as a result of a certain state of destruction of the wood cells, the thicker hyphae become colored dark rust brown (17a), a coloring which I have not observed in the thin hyphae.

How old the mycelial threads can be is a question which must remain unanswered temporarily. The numerous holes in the cell walls, which one observes in the case of a high degree of rot (Fig. 17 & 18), speak for the fact that after a certain time the fungi die and become reabsorbed, therefore disappearing. New hyphae grow through the wood cells so that often one observes fresh uncolored hyphae (Fig. 17g) next to older brown hyphae (17a) in cells where numerous empty holes in the walls (17 c.d.) give proof that already many hyphae have been reabsorbed again. The further the destruction of the wood has progressed, the more readily can be detected the fine colorless hyphae, which arise not only as side branches of the thick hyphae but sometimes directly as continuations of the latter (17h).

These white hyphae contain more or less numerous small granules and in their formation correspond to the white Nachtfaser *Nyctomyces candidus* described by Willkomm (Plate III, Fig. 60). Fusion of several hyphae into broader strands occurs not rarely. If the destruction of the wood has progressed so far that already holes have developed in the wood substance, then also the existing thick brown hyphae lose their color (Fig. 18a), which indeed must be considered as an indication of the beginning of the dissolution of them.

When the hyphae can freely develop in cavities in the tree stem, the mycelium forms either thin loose skins or thick and strong fungus flaps or solid masses of fungus of considerable dimensions.

The rotting of the stem and drying out of the wood very often is linked to a contraction of the latter. Cracks arise in the direction of the medullary rays; very frequently, however, the inner heart separates from the outer mantle of wood so that a concentric cavity forms. This type of crack and cavity is soon filled with abundantly prolific hyphae, as Fig. 1g.g. and 17a illustrate.

There arise thick brown fungus skins (*Xylostroma*) similar to the well known tinder fungus.

Further, between the bark and xylem below the fruiting bodies where these come forth from an unhealed branch stub, by growth in thickness of the tree in the region of the attachment there arises a cavity (1e), which is completely filled by a mass of fungus. Not rarely is the branch stub under an old fruit body completely destroyed by the parasite, and in its place appears a brown mass of fungus which in practice is recognized as a concealed branch fungus. If one places a block of ring-shake Scots pine wood in a moist room for a longer time, for example in a cellar, then the mycelium grows forth over the cut surface and forms a brown mass of fungus on the wood up to a finger-thick. Since, as we will see later, during the destruction of the wood numerous larger and smaller holes arise in the wood, the medullary rays especially often disappear first, thus these offer the mycelium in these holes numerous opportunities for abundant development, whereby it appears first as a brown mass, which later is transformed into a white mass of fungus. In cracks and splits in place of the brown thick mycelial skin layer, a thin loose tissue of white mycelium appears which consists, indeed, in part of the brown hyphae which are involved in the decomposition and in part also from the colorless hyphae, which are living to a certain extent at the expense of the brown fungus mass (Fig. 17b), and which derive their origin from these.

In reference to the mycelium must here be mentioned only that the structure of the hyphae fully agrees with those hyphae which form the substance of the fruit body (Fig. 9, 10, 13, 14), with the sole difference that the latter are more abundantly septate and somewhat thicker.

What has been said indicates that the observation of Caspary, according to which fruit bodies of *Trametes Pini* are supposed to have developed from *Rhizomorpha,* without any doubt is incorrect. The mistake will be caused by the fact that a Scots pine infected by *Trametes Pini* was attacked at the same time by the mycelium of *Agaricus melleus.*

THE FRUIT BODIES AND REPRODUCTIVE ORGANS

Since the sapwood of the tree is always healthy and fungus free thus the mycelium can reach the outside of the tree from the inside only by utilizing not fully healed over ingrown branch stubs as breaks through the sapwood layer.

Fruit bodies therefore never arise at another point of the branches of the stem than where unhealed-over branch stubs or branch holes occur. The mycelium which grows in the wood of the branch (Fig. 1a), just as in the heartwood of the stem, spreads itself from this outward between the bark scales which surround the point of branching externally. The killed tissues of the bark scales are completely grown through by the mycelium, so that

[46]

sometimes these scales are almost entirely transformed into masses of fungi. The brown hyphae in the tissue of the bark scales are very abundantly branched, however, appear much less frequently in comparison to the colorless, white mycelial threads which, partly thick, partly extremely fine, bore through the walls and often completely fill up the interior of the cells. The bark of Scots pine arises from the phloem in this way; cork layers form in the phloem parenchyma, which cut off more or less large scales from the still living inner phloem. These cork layers consist of two layers of very thick walled cells, which are bound to one another or rather are separated from one another by a zone of very thin walled cork cells. The outer layer of thick walled cork cells, which consequently covers the inner surface of each bark scale, is always more strongly developed than the inner layer. Often this consists only of one thick walled cork cell to which are joined towards the interior still a row of starch or chlorophyll bearing cells. The middle layer of the cork layer, which consists of extremely thin and therefore easily ripped apart cells, presents to the mycelium which comes forth from the branch opportunity for abundant development. Between the bark scales the fungus mass forces out in different spots and separates the scales from the surface of the stem (Fig. 2).

The part of the young fruit body which projects freely between the bark scales shows in the beginning a light rust yellow color and a uniform velvet-like surface. The hyphae which end in the surface show for the most part usually cylindrical end-members which are rounded off at the tip (13a), which, with continuous moist air, lengthen themselves by apical growth. Mixed with these hyphae one sees those whose end cells are provided with characteristic hook-shaped formations (Fig. 13b.b.). These arise in cylindrical rounded off end cells in the following manner: the thickening of the cell wall which occurs at the expense of the cell contents is not limited to all sides uniformly but to a thin strip of the primary cell wall (Fig. 13c.c.), while the primary cell wall is reabsorbed. Sometimes the hyphae also end in awl-shaped points (Fig. 13d), indeed also show spore-like abstrictions (Fig. 13e) or cylindrical end cells on long, thin, and thread-like stems (Fig. 13f). If after a rest period renewed growth occurs, thus the ends of the hyphae of the velvety cushion lengthen in a different manner (14). The lengthening by apical growth begins immediately on the uninjured hyphae or the inner cells develop into new slender hyphae (14a.a.) after the end members, especially those which possess hooks, have been broken off.

I must still leave undecided the question whether the breaking off of the ends of the hyphae is only a result of external conditions or whether with the resumption of periodically interrupted growth a pressure emanating from the inner cell results in the breaking off and shedding of the tip.

With the exception of a more or less broad margin which soon appears swollen, the surface of the young fruit body protruding from between the bark scales is covered with small pits, the first rudiments of the canals, and becomes thereby the hymenial surface (Fig. 2).

The pits (1,2,3,5i) are either very small in the beginning and arise then by the pushing apart of the hyphae which stand at right angles to the surface, whereby a funnel-shaped hole forms, or they arise immediately as broad-based, non-funnel-shaped pits in the following manner: the hyphae lose here and there their longitudinal growth, while in the interstices between these spots the continuous growth of the hyphae throws into bold relief the walls of the canals. The attainable width of the canal is dictated by the distance of the pits which arise in the juvenile fruit bodies. A canal obtains the width which it has attained in its youth for all following time (5d). Its length on the contrary increases constantly, by the fact that the more or less parallel running hyphae of the walls periodically lengthen themselves at the tip (8a). With rare exceptions (7) the canals always run completely perpendicular (5); as a result of this in juvenile fruit bodies the mouths of these obtain a very stretched shape. With the growth of the walls of the canal there occurs simultaneously an enlargement of the fruit body by a periodically interrupted growth of the swollen margin in the vicinity of the hymenial surface. The swollen margin on the upper side of the hymenial surface (3b, 5b) is significantly thicker than on the remaining sides and is distinguished above all by the orientation of growth of its hyphae. Enlarging themselves in a more or less radial horizontal direction, the hyphae of the swollen margin form the upper sterile surface of the bracket-shaped fruit body, whose slanted underside, up to a very narrow medullary body which lies against the bark of the tree, is occupied entirely by the hymenophore. The hyphae of the swollen margin in the upper half bend themselves bowlike towards the upper surface (5b), in the lower half towards the lower surface ($5a^x$), while in the middle the enlargement of the swelling occurs continuously. The light rust yellow color of the velvety swelling very soon is transformed into a darker, almost brown color. The velvety sheen disappears especially in the upper sterile remaining part of the new swelling due to the influences of the atmosphere. The hyphal ends collapse, later weather away, and give the upper surface a rough black appearance. Since growth is periodically interrupted and according to weather conditions is more vigorous or more weak, thus zones form on the sterile upper surface of the fruit body, which are broken through repeatedly by radially running cracks (1,4,5). The lower part of the swelling serves for the enlargement of the hymenial surface. Not far from the margin new pits arise (3i, 5i), which, with the general growth of the fruit body, also become canals. While the upper swollen margin of the fruit body causes the enlargement of the organ in the previously described manner, the lateral borders likewise expand by growth of the hyphae of the swollen margin, thereby, however, always remaining tightly appressed to the bark of the tree, also they always develop considerably thinner, so that they are scarcely detectable (3a.a.). With the enlargement of the hymenial surface there also occurs a corresponding increase of the pits and canals on the entire circumference.

Usually on one branch stub there occurs only a single fruit body; sometimes, however, on different sides of the branch there arise simultaneously several so-called "Schwamm."

If a fruit body is broken off, thus as a rule there arises in the same spot several new fruit bodies, since the residue is immediately capable of vigorous growth.

Very often an age of 50 to 60 years can be proven with certainty for a fruit body (Fig. 1).

Whether this age can be still considerably exceeded is doubtful to me.

The growth of the "Schwamm" finally ceases, either in all of its parts or only partially. The hymenial surface grows no more, begins to weather away and to become rough (6a.a.), the entire fruiting body becomes crumbly and falls off. Very often a regeneration occurs on already partially dead fruit bodies. From the canals grows forth a new cushion (6u, 7c.c.) in which completely new pits and canals are able to develop (6d).

The formation of a new cushion is to be attributed to the same hyphae which regularly cause the plugging up of the canals after these have fructified for a series of years, as will be described in the following.

On older fruit bodies one always notices that the canals or tubes often are not open for their entire length, but the greatest part ultimately are plugged up by a special filling so that only the youngest part stands open (5c). If one investigates this youngest part (8), thus one notices that the hyphae of the walls (8b, 9, 10) run rather parallel and by the mouth of the tube (8a), where very often the periodic growth of it appears clearly due to the coloration, still no basidia or hairs have ramified in the cavity of the canal.

Only somewhat deeper in the interior of the tube, hence perhaps after several years of age, the club-shaped swollen basidia, which form the hymenial layer, penetrate from the wall on longer and shorter branches of the parallel running hyphae of the hymenophore (9b). A relatively not very large number of basidia develop four often very long sterigma which produce round egg-shaped spores on their ends. Many basidia appear to always remain sterile (9). Scattered throughout the hymenial layer and growing forth simultaneously with the basidia arise very strikingly formed filaments often detectable with the unaided eye. They originate from thick hyphae which approach the hymenial layer at acute angles from within and the initially colorless, later dark-brown, thick walled awl-shaped filaments penetrate into the cavity at nearly right angles to the hymenial surface (9a.a.). If, depending on weather conditions, growth of the fruit body begins in August or September, then new basidia and setae grow throughout the hymenial layer of the first year (9), so that the initially very thin layer thickens considerably, the cavity of the tubes becomes constricted.

The first formed basidia and setae appear no longer or only with their tips on the surface of the canals (Fig. 10), the hymenial layer attains in the course of time a considerable thickness. If one takes in September a section from the bottom of a still open portion of the tube thus will one be able to observe besides old setae also those newly arising in all stages of development (10a^x, a^{xx}, a^{xxx}). The hyphal branches growing forth between the old basidia form on their tips new basidia, attain at times an unusual length so that they penetrate further into the cavity of the canal. Some of these, however, do not appear to reach normal formation of basidia; they cut off on their ends larger or smaller roundish spores, branch, and frequently show various abnormal formations, by which sometimes it is doubtful whether these do not belong to foreign fungus species which have penetrated into these canals (Fig. 10). In the same degree that the walls of the canals lengthen themselves through growth of the hyphae (8a), the interior of these fill themselves with a mass consisting of hyphae matted together with one another (c).

This filling material consists of hyphae which grow forth from the hymenophore over the hymenial layer (10e.e.), and branching repeatedly and growing through one another plug up completely the interior of the canal. Also this plugging up occurs periodically. There arises at the bottom of the canal on the ends of many of the plugging hyphae exactly such awl-shaped setae (8e.e.) as come forth between the basidia of the hymenial layer. On a smooth cross section (5) the filling is immediately distinguished from the walls by the absence of the silky luster.

The spores (11), which are formed mostly in fours on the basidia, are initially colorless, however at greater age take on the brown color which is characteristic of the older hyphae.

They are broadly egg-shaped, rarely almost spherical. Their length reaches about 5.3 microns, their width about 4.5 microns.

In very many spores appears a very sharply outlined sphere, which, indeed, must be considered as a drop of oil.

The thick wall appears especially clearly in such spores. The greater part of the spores appear never to leave the canals. Some even germinate in them (12).

The end of the germ tube was broken off in the case of the illustrated germinated spore which originated from the interior of a canal. I was not successful in bringing the spores to germination artificially. Frequently one finds large clumps of ungerminated brown colored spores embedded in the filling material in the older, grown-together part of the canals.

The annual growth occurs from the middle of August or September to the beginning or end of November and is influenced by a certain moisture condition of the fruit body, which begins and maintains itself only with continuous air moisture.

THE MODE OF LIFE OF THE PARASITE

Since the development of the fungus mycelium as a rule can occur only in the heartwood of Scots pine, thus the youngest age classes are exempt from the disease. Only with the formation of heartwood is the tree exposed to the danger of infection. Numerous infection studies on thirty- to forty-year-old Scots pine were completely unsuccessful, while, indeed infection in older Scots pine succeeded without exception.

The infection was carried out in the following manner; an increment core of mycelium-containing Scots pine wood was inserted into a bore hole produced with the same increment borer (Presslers). The latter was immediately sealed externally with grafting wax. It is self-evident that on the bored out core it was confirmed with the greatest certainty whether the tree was still healthy at the time of the infection.

Within a year the wood in the vicinity of the infection spot and sometimes up to one decimeter removed showed the condition which is illustrated in Plate III, Fig. 16.

Numerous mycelial threads, whose origin from the mycelium of the diseased wood allows no doubt, occurred in the interior of the wood fibers, especially developed in the medullary rays, expanding in which they spread the disease rapidly into the interior of the tree.

I have encountered trees 60 to 70 years old with the fungus, although only rarely so that perhaps the 40 to 50 year age is to be designated as the one in which pine becomes susceptible to the infection that occurs.

Since the sapwood for a width of several centimeters remains white and fungus free (d) in such stems which in their interior show a high degree of disease, since the latter never emanates from the roots as does the red rot of spruce, but in most cases begins in rather highly situated stem parts in the upper or lower half of the crown, sometimes restricted to only a few branches of the crown, thus there are only the broken-off or cut-off larger, that is to say, heart-containing branches whose stubs, leading through the sapwood layer as natural bridges, make possible the penetration of the parasite from outside into the heart of the tree (Plate III, Fig. 1a). From the infection site the disease spreads out in all directions, with special preference to the longitudinal direction of the stem, which is often very clearly detectable on cut wood, boards, or beams, which show long red brown streaks in the rest of the at least apparently still healthy wood.

Very frequently a tree affords still very worthwhile commercial timbers if one or several blocks are cut off for fire wood. In other cases, however, decay appears through the entire stem and even penetrates into the roots. The progress of the decay process is not always the same. As a rule it is as follows: The mycelial threads, spreading in the medullary ray cells and in the wood fibers, bore through the walls in numerous spots by which they send out quite short or also longer side hyphae whose thickness is not always the same (16). The pits are selected only occasionally as penetration spots; often a mycelial thread branches in the lense-shaped pit cavity (16b), and bores through its walls in several points.

The size of the holes which are bored through by the hypha in the walls, correspond to the diameter of the hypha and in all parts are equally wide, that is to say cylindrical (17d).

The wood fibers at first suffer no noticeable change either in their thickness or in chemical composition. On the contrary, often turpentine is secreted in them drop by drop on the walls. Whether this has flowed out only from the destroyed resin canals and has precipitated then in the interior of the wood cells or whether this is also a decomposition product of the contents of the medullary ray cells and so forth must remain undecided.

A further stage of the disease (17) shows the walls of the wood cells bored through (17c,d), the hyphae belonging to these, however, already have been reabsorbed.

Dark brown colored older and light colored younger hyphae continue boring through the walls which show, indeed, a light browning but no noticeable thinning. The holes which are present in these are not noticeably changed in size.

The proportion of the thick hyphae to the fine (17e) has improved somewhat in favor of the latter. The direct connection of the two types of hyphae can be detected with doubtless certainty.

Up to this developmental stage the progress of the disease is indeed always the same. This type of wood is still strong and only by the dark red coloring distinguishable from healthy heartwood. The further progress of the destruction in different trees and in different spots of the same tree is different. As a rule one sees arise especially in the thick summer wood layer, but also in the spring wood layer, holes at first small, later increasing in size (1b, 15), whose walls are covered by silver white shiny threads, by which the holes appear very strikingly as white flecks in red wood. The surrounding wood tissue becomes strikingly brittle and full of holes and is very easily broken up during manufacturing of thin sections. This stage (18) is of the greatest interest in that it affords the best conclusions about the process of decomposition. The brittle wood shows its fiber walls still in their original thickness. The holes which are produced by the fungus hyphae and the pits on the other hand are enlarged to a striking extent, their shape in a cross-section of the wall (18d) is no longer a cylindrical form but is strongly bulged out within the cellulose layer.

Apparently the inner cell wall layer affords the greatest resistance to the destruction, also the joint primary wall is not destroyed as easily as the so-called thickening or cellulose layer. The innermost layer of the wall protects it against destruction, which with the greatest consequences has begun from the margins of the bore holes outward.

Relatively few thicker and finer mycelial threads are detectable still in the fibers; however, these no longer colored brown but extremely transparent must indeed be already exposed to the process of dissolution.

If one makes a section through the margin of one of the white walled holes in the wood, the cellulose and primary wall disappear without gradual transformation and only the innermost layer of the fiber remains (18b.ff). It can scarcely be doubted that also a chemical change precedes the complete dissolution of this layer, which perhaps is indicated by the uncommonly strong and rapid swelling of the substance with treatment with sulfuric acid.

The extremely thin inner membrane is completely colorless, and consequently to the unaided eye the covering of the holes appears white colored. The individual fibers are no longer bound to one another but stand completely isolated. Boreholes, pits, and fungus hyphae are still detectable only with the most favorable lighting. Only the rings of pit cavities already mentioned by Willkomm in the Rothfäule der Fichte withstand the decomposition even if, to be sure, in most cases they are split into many pieces.

The tender colorless inner membranes maintain to the end a very fine granulation (18h), by which the final stage of dissolution of this layer also becomes pointed out. The medullary ray cells undergo the same

[52]

transformations, as was described before for the wood fibers, and finally only the resin perhaps present in the cells (18e) resists the decomposition. The inner lumen of the holes which arise thus contains as the last remains of the destroyed wood only a yellowish white dust with some fungus hyphae (19). These are particles of resin from the medullary ray cells or from the resin canals.

The process of destruction of the wood does not always proceed in the previously described manner, often the formation of the white flecks in the wood occurs not at all or only in spots.

The medullary rays of the summer wood are the ones which are first completely destroyed by the especially abundantly developing mycelium in them, and from these outward the destruction proceeds first to the summer wood layer and later also from here to the spring wood layer.

While the latter to the external appearance especially by its luster seems to be little changed, and only by testing with a knife by its weakness is its degree of decomposition detectable, the entire summer wood layer is already consumed and lusterless, however not colored white but yellowish.

Microscopic investigation shows that with this type of destruction process the walls of the wood either crumble into white irregular pieces (17f) which arise through splits and cracks (17i) in any direction, or that the walls have numerous longitudinal cracks (18k) whereby they crumble into uniformly broken up very small longitudinal pieces. Also with this process of destruction the cellulose layer disappears earlier than the inner boundary layer of the walls; nevertheless only exceptionally does a white coloring of the substance appear before complete dissolution, while the destruction occurs more uniformly through the entire summer layer of the ring and the cavities that arise fill themselves up with the brownish yellow broken pieces of the walls, the fungus hyphae and the particles of resin as a porous felt-like mass.

The development of fungus hyphae in the interior of the wood fibers often is so abundant that the entire cavity is filled thereby (17f). Especially in the summer wood the resin canals and their neighbors are destroyed early, the fungus mycelium which penetrates from these into the medullary rays and destroys these even earlier than the surrounding tissue must perhaps explain the phenomena that on the one hand in the summer wood, where the resin canals are more frequent than in the spring wood, the destruction proceeds earlier and more rapidly, that on the other hand this does not hasten uniformly rapidly through the entire tissue but does so in spots. From the resin canals which in the tree run parallel to the wood fibers, the medullary rays, which by their more horizontally running resin canals are in direct connection with these, are drawn first into the destruction.

From these the destruction spreads out on all sides. Instead of the medullary rays a strand of brownish or white mycelial threads often appears, as generally all cavities afford the mycelium the opportunities for more abundant development. As a result of the shrinking together of the xylem, which is linked with the decay and the loss of water from it, there form not only radially running cracks, but also very often the outermost ring layers loosen themselves like a mantle from a thicker or thinner heart. There arise in this manner ring cracks (Ringspalten), which, indeed, has given rise to the name of ring-shake. The splits and cracks which arise by the shrinking of the wood are covered with a thick or thin membrane of rust brown fungus hyphae which especially grow forth from the medullary rays. Often wide cracks are filled with a mass of fungus (1g.g.), which show the same composition as the substance of the fruit body (17 right). In later stages of development this mass of mycelium disappears just like the mycelial threads in the interior of the fibers, in that they themselves fall to the decomposition. This is initiated in the following manner; fine, colorless hyphae, which rise from the brown thick hyphae (Fig. 17a^{x}), form an extremely delicate white mold-like covering over the brown mycelial layer, which finally, after the brown mycelium has disappeared, is also again destroyed.

In the described way, and often in different manners there arise in the heartwood of the tree continually larger and more numerous holes until a complete hollowness occurs and the wall of the shell of the tree which remains behind is covered only by a yellowish meal (particles of resin). Even before the interior of the main stem shows a higher degree of destruction, this can already be seen on any enclosed branch stub (1a). These form not only the only bridge by which the parasite penetrates from outside into the interior of the tree but also in the reverse can lead to the development of its fruit body, the so-called Schwamm. This has already been described in detail. Worthy of mention in addition is the circumstance that on the border between the healthy sapwood and the fungus-infected heartwood lies a fungus-free zone which is very heavily resin impregnated. Likewise, in the vicinity of the branch stub on which the fruit body comes forth the wood is completely resin impregnated and dark colored (1). This supports the supposition to ascribe the turpentine content of the xylem to the fact that with the destruction of the resin canals in the interior of the tree, a secretion of resin towards the outside occurs through the radially running resin canals.

The opponents of the view that diseases of plants can be caused by parasitic fungi will not fail to attribute the bark-shake or red rot of Scots pine to other causes. Great age of the tree and the loss in function of heartwood which thereby occurs, poor or too favorable soil conditions, unfavorable weather conditions due to which the wood of an annual ring has not been able to reach maturity, and other unfounded suppositions will be too widely loved by most unscientific practitioners as explanations of red rot, to abandon their prejudice against the results of microscopic research and scientific methods of investigation of the forest. Scientifically based forestry practitioners may still, for settlement of each alleged opinion, object that great age of the wood cannot be the cause of the decay when it often occurs on 60-year-old trees while it is not always present on 200-year-old trees; further, that the oldest wood of the tree is not the first to become diseased but that the disease often begins on the branches of the tree crown.

Loss in function of the heartwood likewise is not the cause since on the one hand the disease must occur quite commonly with a certain age, on the other hand the heart of Scots pine is not completely functionless. Concerning the question to be answered whether the heart of Scots pine is still sap conducting, in the spring of 1871 I had the xylem of three 110-year-old Scots pines of about 40 cm diameter sawn through in a ring fashion about 20 cm above the soil sufficiently deep that stem I was cut 13 cm deep, thus it contained 40-26 = 14 cm of unsevered heart. Stem II was cut 10 cm deep, contained 38-20 = 18 cm of heart. Stem III was cut 7 cm deep, contained 40-14 = 26 cm of heart. By shoving an oiled paper into the sawcut the possible mechanical absorption of moisture from the lower stem portion was counteracted.

In all three stems the sapwood was completely severed, in the case of Stem I, in comparison with the tree, only an extremely small area of the innermost heart was left behind for sap conduction.

The result of the experiment was that Stem II in a fully green condition was thrown by a wind storm on 21 September of that year; Stem I maintained itself green through the summers of 1871 and 1872 and was broken off by wind in August of 1872.

Stem III in a still completely green needled condition was felled by me in September of 1872 to be able to undertake thereby further investigations on the water content of the wood. The proof of the sap conductibility of Scots pine heartwood thereby must have been supplied sufficiently, although it still will not be disputed that the sapwood under normal circumstances preferentially conducts the sap. Therefore, it cannot be considered as the loss in function of the heartwood fibers which, so to speak, brings on the rot as a natural result.

According to my observations up to now likewise there can be detected in soil quality no sufficient factor which can serve as an explanation of the disease. Fungus infected trees have been found by me on soils of the poorest and best quality, on dry, fresh, and, indeed, also swampy soils. Since unfortunately the gentlemen of the green forest, in regard to my request expressed at the Swinemünder Forestry Congress of this year to submit to me their observations about the occurrence of barkshake in respect to soil quality, have come forth to only an extremely limited manner, thus I stand in this direction still completely isolated to my narrow personal experiences. According to these, this disease, as already said, occurs on all soil types, most frequently, however, on the best soil classes, especially on soils which likewise can be allotted to beech.

If one still considers that the disease always occurs especially severely where the stand, because of its exposure, is exposed to the attack of storms, that trees which in their early years have suffered numerous broken branches are especially frequently red rotted, that finally in a stand whose location in the vicinity of a village has had the result that in a criminal manner severe debranching was undertaken, it is the case that almost every tree had become diseased in its interior, thus all of these observations can be explained quite well by the fact that in these cases infection by the spores of *Trametes Pini* had been offered numerous opportunities by the freshly broken branches. To me it is very doubtful that branches which die by natural suppression make infection possible.

If the suppressed twigs and branches fall in a dry condition, thus there remains, indeed, under certain circumstances small branch stubs or cavities are formed in the tree which are walled over only after several years. The end of the branch stub which remains behind in the stem may itself be still so short, that it probably must always find itself in the condition that the development of a young *Trametes* plant from disseminated spores cannot occur. The heart of the branch stub is already resin impregnated before natural abscission of the branch, the sapwood is severely destroyed by other fungi growing saprophytically only in the dead wood, the tip of the branch stub is generally decayed and dried up.

If the tree also protects itself against infection in the following manner, that the suppressed twigs and branches die, dry up, become resin impregnated, and are destroyed by other saprophytic fungi before abscission, thus this explains the fact not all trees become red-rotted in the course of time, that in the cited cases the disease occurs especially frequently and severely.

In the vicinity of towns and cities where, due to cutting off and breaking off of green branches, numerous fresh branch wounds are formed, fungus infected trees are very widely distributed, as for example near Neustadt on good and poor soils.

On the freely lying stand borders, on hilltops and upper slopes exposed to the winds, due to windstorms more frequent broken branches occur than in protected sites, therefore also fungus infected trees occur more frequently on such sites.

On the best soil classes with very vigorous growth and heavy crown development, broken branches must likewise occur more frequently with strong winds than on the lower soil classes on which the narrow-ringed wood of the branches is much stronger and more resistant than the broad-ringed branchwood of more rapidly growing trees.

It still should be mentioned that the emergence of fruit bodies on diseased trees can occur only on such branch stubs where walling over has still not yet occurred. The part of the branch stub which is buried in the wood, as long as it is still living, is destroyed by mycelium of *Trametes* and the latter can then in its abundant development through the crumbly decayed portion of the branchwood not be hindered in its pushing outward towards the outside.

If now in conclusion we ask whether, according to the results of the investigations which have been obtained, measures can be applied against the disease in forestry management, the question likewise is to be answered in the affirmative.

In the first place it is to be strongly guarded that the criminal breaking off and cutting off of branches be eliminated. The pruning of live branches, which has recently been accepted in forestry practice for the purpose of the production of knot-free commercial lumber appears to me on the contrary for the present to be still permitted if it is employed in young pole timber up to 30 years of age, since on the one hand the branch stubs still contain no heartwood, since on the other hand the infection is less threatening in the interior of larger stands of young wood far removed from old fungus infected trees.

However, it will be an item of further investigation and experimentation to be able to answer this question with greater certainty than is now possible for me at this time.

The cutting out of fungus infected trees in thinnings and so-called salvage cuttings are definitely to be employed, to be sure on two grounds. The fruiting body which releases the spores is thereby removed, which, even if a breaking off of this was practical to carry out, would form a new one in a few years. With as much as possible complete removal of the infected material, self-evidently the danger of the disease is reduced. The fungus infected trees at early cutting times in many cases still produce worthwhile pieces of useable lumber, likewise also more combustible wood than when they remain standing until normal cutting time. Whatever apparent growth in mass they put on as a rule is considerably less than the loss in wood which is brought about simultaneously in the interior of the tree by the disease which spreads out from the infection site. The measure is feasible since with annual cutting the stems of the fungus infected trees which have died are easily found and can be worked up with the remaining stems.

If a stand is so severely diseased that by the removal of fungus infected trees considerable holes will arise in the stand, thus it already serves the financial interest of the woodland owners to carry out as rapidly as possible the cutting of the stand since the value of the wood will steadily decrease.

Explanation of Illustrations

(Plate III, Figures 1-19)

Fig. 1. Stem cross-section cut from a stem of a red-rotted Scots pine with an ingrown old branch connection (a). The heartwood (b) is red colored and shows numerous flecks and holes distributed partly equally, partly concentrated in narrow and broad zones. On the border between the healthy white sapwood (d) and the corroded heartwood occurs a red-brown still strong zone (c) which towards the inside is very fungus-rich, towards the outside fungus-free but strongly impregnated with resin. Ring-shaped cavities arising by the shrinking of the diseased wood are filled with a brown, tinder-like mycelial mass (g.g.). Between the bark and the wood which surrounds the branch stubs as a result of the ceasing of growth in thickness a cavity has formed (e) which is completely filled with a brown fungus mycelium. On the sectioned fruit body of *Trametes Pini* one can detect a part of the here exceptionally oblique sterile upperside with concentric deep furrows and longitudinal cracks, as well as a section through the interior. The tubes are empty only on the lower-most portion. The age of this is determined by the number of the annual rings, which are absent in the cavity (e), up to 50 years.

Fig. 2. A young fruit body from a Scots pine branch. The bark scales in the vicinity of the still not overwalled branch stub are forced apart from one another by the mycelium springing out as it were between them and lifted off from the substratum. The freely projecting surface shows the first developmental stage of the tubes in the form of numerous small pits.

[57]

Fig. 3. Older developmental stage of a fruit body. The swelling in the vicinity of the hymenial surface is on the upper margin (b) very thick, on the remaining sides (a.a.) only narrow. The tubes, which are always reproducing themselves anew on the periphery as small pits (i), deepen by an almost perpendicular lengthening of the hyphae of the wall, as a result of which with a very sterile orientation of the hymenial surfaces the openings of the tubes take on an elongated form.

Fig. 4. An older bracket-shaped fruit body in side view.

Fig. 5. A similar fruit body in cross-section. The swollen margin which enlarges itself annually in the region of the hymenial surface forms the so-called medullary layer (a.a.). The hyphae bow themselves partly toward the hymenial surface and produce on their upper surface new pits (i), partly bow themselves towards the outside and end in the soon weathered away sterile upper side (b.b.^{v}) or adhere tightly to the bark scales of the tree. The space between the medullary layer, not unsimilar to a cone with a spherical base, is occupied by the hymenophore, whose tubes (d) run perpendicularly, and after a few years are filled by hyphae which arise from the walls, so that only the youngest part is open and producing spores.

Figs. 6 and 7. A very old fruit body which as Fig. 7 shows in cross-section, has not formed in a normal manner but which at some time earlier has been severely damaged, so that a completely new hymenial layer has formed itself on the part remaining on the tree. The orientation of the tubes as a result of this is not perpendicular but oblique and not everywhere the same. The hymenial surface is here and there completely destroyed, weathered away or eaten by insects (a.a.). Here and there it is formed normally (b.b.) with open canals. The greater part of the surface shows velvety cushions (c.c.) which arise in the following manner; abundant development of mycelium has occurred from the interior of the canals outward. On the surface of the cushions arise here and there further new pits (d) whereby a regeneration of the fruit body is brought about.

Fig. 8. Longitudinal section through the youngest part of a tube (a). The youngest end of the tube wall whose growth is periodically interrupted is distinguished by the lighter coloring of the younger parts. The hymenial bearing wall (hymenophore) consists of more or less parallel running hyphae (b), from which side branches later grow out into the tube and completely plug it (c). The surface of the wall is covered with numerous long awl-shaped hairs and basidia (d.d.).

Fig. 9. A part of the hymenial layer and the hymenophore. Fig. 8 at x enlarged. The wall hyphae come forth obliquely below the hymenial surface and widen themselves into awl-shaped hairs (a). From other hyphae side branches arise at right angle to the surface of the wall, which on their end swell into club-shaped basidia (b). These bear in part four sterigma with spores in different stages of development (c.c.).

Fig. 10. A part taken deeper in the interior of the tube at the time of the growth of the fruit body. The hymenial layer has become thicker in the course of the years, in that annually between the old basidia (c.c.), new ones grow out (b.b.) which in part project on long hyphal branches into the tube. The significance of some abnormal formations (d.d.) is unknown. The oldest hairs project in part only with the point below the hymenial layer. New hairs (a^{x}, a^{xx}, a^{xxx}) are nascent. Individually and in groups (e.e.) thick hyphae which become the matted filling material of the tube grow forth from the wall. These hyphae also develop awl-shaped hairs in the base of the tube before the end of each growth period.

Fig. 11. Mature spores originally colorless but at greater age brown colored.

Fig. 12. Germinated spore from the interior of a tube with a ruptured end.

Fig. 13. The ends of hyphae from the velvety cushions, which are formed not only on the margin of the fruit body (Fig. 3b) but also on the end of the tube walls in a resting condition. The end cells mostly are simple cylindrical, rounded-off on the point (a.a.) partly seen with characteristic little-hooks (b.b.) which at (c.c.) are still developing. Other hyphae end awl-shaped (d) or bear round spore-like swellings (e). Sometimes cylindrical cells are borne on long thread-like hyphal ends (f.f.). The surfaces of individual hyphae are covered with granules (crystals?) (d).

Fig. 14. The ends of hyphae characteristic of growing velvety cushions. From the broken-off ends the inner tube (a.a.) arises as a new thin-walled hyphae (a.b.). Other hyphae lengthen themselves directly.

Fig. 15. A piece of red-rotted Scots pine wood with numerous holes whose walls are white-colored.

Fig. 16. Radial section through Scots pine wood which has been infected artificially a year previously. The mycelial threads develop abundantly especially in the medullary rays, still bear plasma, are colorless and bore through the walls. Individual fine hyphae (c.c.) arise from the 3 micron thick main hyphae.

Fig. 17. Red-rotted Scots pine wood. Numerous hyphae have already disappeared which is detectable by the numerous holes in the wall (c.d.). Brown-colored older hyphae are found next to colorless young hyphae (g). Fine hyphae (e.e.) also arise here from thicker hyphae. At the right a very abundant development of the fungus has occurred, as a result of which the fiber walls are largely destroyed and broken into pieces (i.f.). At (a.a.) in a hole which arose by a shrinking due to drying, the wall is covered with a brown mycelial layer whose hyphae in a further stage of destruction are again colorless (b.b.) and are replaced by thin hyphae.

Fig. 18. Red-rotted Scots pine wood (Fig. 15^x). The still brown-colored brittle xylem shows cell walls bored through like a sieve. The bore holes and pits are enlarged, in cross-section (d.d.) within the thickening layer of each wall they widen out convexly. In the vicinity of the holes the cellulose layer and the primary wall disappear suddenly (b), and only the inner border layers of each fiber (ff) remain behind. At this stage almost only very fine colorless fungal hyphae still occur (g). Complete dissolving appears at (h), where the thin border layer itself crumbles away into the smallest molecules. (i) shows the inner empty cavity of the hole. (k) shows a likewise frequent condition of dissolution of the cell wall. The little resin particles in the medullary rays (e.e.) of the resin canals or xylem fibers alone withstand the destruction.

Fig. 19. The yellowish powder which remains as the end product of the decay process and consists only of resin and some fungus hyphae.

Trametes radiciperda R. Hrtg.

(Plate III, Fig. 20-29)

In the investigation of trees killed by resin flux, I found many plants suddenly killed in their most vigorous growth in which not only the characteristic mycelium described above as *Rhizomorpha fragilus* was absent, but also a resin secretion was not noticeable at all or only in very small amounts on the root stock. Only the sudden dying often in the middle of the development of new shoots agreed with the previous disease.

I have observed this disease many times in the vicinity of Neustadt-Eberswalde on young 5-20 year old Scots pine, on juniper, and also on different hardwoods and in each case found on the roots or on the root stock a fungus which, by its occurrence under the soil, seems until now to have avoided the eyes of the mycologist.

Since I hold this fungus to be the cause of the sudden dying, I give here its description.

THE MYCELIUM

The mycelium of the fungus which develops in the phloem and xylem of attacked plants has great similarity to *Trametes Pini*.

It is, however, always colorless, although under some circumstances it takes on a brown coloration.

The hyphae are very sparsely septate (Plate III, Fig. 28, 29) and abundantly branched. From the thicker hyphae, which reach up to 2 microns in diameter, arise finer hyphae as in the case of *T. Pini* (Plate III, Fig. 29).

The walls appear double layered only if the hyphae are empty.

The thick and thin hyphae bore through the walls of the xylem and phloem organs with the greatest of ease, often form in cavities of the killed phloem a loosely matted fungus mass, which penetrates forth from the bark cracks of the roots and forms irregularly shaped, yellowish-white little bumps, usually incrusted by sand grains, from which here and there fruit bodies of the fungus arise (21a).

THE FRUIT BODIES AND REPRODUCTIVE ORGANS

The fruit bodies arise preferentially on the root stocks of attacked plants (20 a.b.) where the humus, moss, or grass covering of the soil favors the development by conservation of a certain moisture level. Often one sees them also on the roots and, indeed, also to a depth to 1-2 decimeters on the side roots somewhat removed from the root stock. They seldom arise individually, usually are clustered (20a) and then form grape-like masses, yellowish-white or snow-white in color.

The mycelium which comes forth between the bark cracks appears externally in the form of small balls of the size of pin heads (20a) individually or united in groups. These then often fuse into a single fruit body (21b) on whose surface, with the exception of the strongly swollen margin, the hymenium-bearing tubes arise in the same manner as has been described in *T. Pini*. The ends of the hyphae of the cushions show the same numerous differences of form (27) which have been described for *T. Pini* (13). The sterile upper surface is distinguished by a gray color from the snow-white remaining portion of the fruit body; moreover, it is distinguished by the presence of a type of rind.

Above the thickly matted white tissue of the medullary body (27a) lies a layer of tangled, dark-colored hyphae (27 c.c.), through which the previously mentioned hyphae (27b) grow at right angles to the interior. The growth of the young fruit bodies is essentially differentiated from that of *T. Pini* in that the swollen margin in the entire circumference of the hymenial surface is strongly symmetrical and enlarges itself centrifugally nestled against the bark of the plant. After the surface reaches a certain size (21c), growth is limited only to the hymenophore. The canals lengthen themselves by growth of the walls at their tips. The bracket shape which is so characteristic of most species of *Trametes* and *Polyporus* appears not to occur at all with this fungus. With several-year-old fruit bodies the hymenial surface in part dies (22, 23).

A new growth of the same type occurs in the same manner as in the case of *Trametes Pini* (6, 7).

The inner structure is completely different from that of *Trametes Pini*. The hyphae do not run parallel but are interwoven in all directions and tangled under one another (24, 25), without predominating in any one direction. The branching is quite abundant, the septations sparse. The hymenial surface, which covers the walls of the canals, is very slightly delimited against the tissue of the hymenophore. These hyphae arise deep in the interior at right angles to the surface of the walls; on their ends they usually thicken club shaped and become basidia (25a). However, only a small part of them form sterigma and spores. The latter (26) are white, somewhat smaller than the spores of *T. Pini*, of less constant size and form. Formation of setae is completely absent; on the contrary, numerous hyphae grow across the hymenial layer outward into the interior of the canal, which, in association with hyphae growing out later, form a filling material which plugs up the older parts of the tube to a degree (22) as is the case with the fruit bodies of *T. Pini*. How great an age can be reached by these perennial fungi must be left undetermined for the present.

[61]

THE MODE OF LIFE OF THE PARASITE

About the mode of life of the fungus only a little can be said at this time. Scots pines, which were investigated very soon after their death, indicated that the entire xylem and bark of the roots had been attacked by the mycelium. In the fibers (29) I detected not only the mycelial threads of various widths but also the holes which had been bored by these in the walls. The killing effect however depends upon the destruction of the phloem and bast (28) since with death of these sap uptake from the roots must cease. The circumstance that a plant suddenly dies which until now has been making its most vigorous growth, speaks for a very rapid spread of the mycelium in the roots; the phenomenon that in the vicinity of a dead plant in the next years the neighboring plants suffer the very same disease indicates the infectious character of this disease. In hardwoods as well as in coniferous forests rather large holes are known to me in which the plants have died little by little, already showing the described fungus.

Whether infection occurs either through spread of the mycelium in the soil or by spores is still unknown to me. Pulling up and rooting out the infected plants appears advisable in any case.

Explanations of Illustrations

(Plate III, Fig. 20-29)

Fig. 20. Root stock of a young Scots pine with a vigorously developed fruit body *Trametes radiciperda*. a. Young fruit bodies clustered together grape-like; b. older fruit bodies.

Fig. 21. Different stages of development of fruit bodies. a. Snow-white young fruit bodies, single or clustered together; b. from the fusion of several fruit bodies with the resulting hymenial surface; c. fruit body (double size) with a thick swollen margin.

Figs. 22 and 23. Several-year-old fruit body in frontal view and cross-section, whose hymenial surface at a.a. is already sterile and decomposed, whose tubes in other spots (b.b.) are still open and forming spores.

Fig. 24. The open part of a tube in cross-section. At a.a. the growing ends of the walls. At b.b. the hymenophore which consists of numerous cross entwined hyphae. At c. the tissue which fills up the pores. At d. the hymenial surface.

Fig. 25. A part of the hymenophore and the hymenial surface enlarged in cross-section. The basidia, which are generally thickened club-shaped at the end (a.), arise from the interior from the matted tissue of the wall. The hyphae are somewhat thicker than the thickest mycelial threads, white, branched, and sparsely septate.

Fig. 26. Spores.

Fig. 27. Cross-section through the sterile part of the swollen margin of a fruit body. From the fungus tissue (a.a.) hyphae arise towards the outside with cylindrical, hooked, or ball-shaped end members (b.b.). The fungus tissue is covered externally by a dark colored rind (c.c.).

Fig. 28. Phloem of a root of a Scots pine which has been killed by *T. radiciperda*. The mycelium consists of both thick and thin hyphae and develops preferentially in the phloem parenchyma whose cell walls are bored through.

Fig. 29. Root fibers of Scots pine with mycelium of *T. radiciperda*.

Aecidium (Peridermium) Pini Pers.

(Plate IV)

The Pine Blister Rust (Kiefernblasenrost)

Causal organism of Scots pine needle rust, canker, rust
or scab of Scots pine and resin top (Kienzopf).

Aecidium Pini is one of the most well known rust fungi of conifers; the symptoms of the disease which is caused by this parasite are so frequently encountered by workers in the forest that it has claimed their attention to an outstanding degree. Since an exact scientific investigation neither of the fungus nor of its disease has occurred until now, indeed the latter has been associated only in part with the former, thus the following results of investigations must be suitable to fill up an essential hole in the knowledge of the diseases of Scots pine. For the orientation of the mycologically non-educated portion of my respected readers, a few remarks about the rust fungi must be presented first.

GENERALITIES ABOUT THE RUST FUNGI

The rust fungi (Uredineae), to which the Scots pine blister rust belongs according to the interesting investigations of de Bary and others possess a distinct alternation of generations.

They occur mostly in three different forms, each of which is distinguished by a special course of development of its fruiting bodies.

At certain times so-called survival spores (teliospores) are produced by the rust fungi, thick walled spores, which, produced individually or one over the other in pairs, do not fall from their supports, but winter usually on the plants on which the parasite has developed (*Puccinia, Melampsora*) and germinate in the spring. The survival spores then produce a thick germ tube (promycelium) which, soon after it has ended growth in length, sends up one to four awl-shaped branches, each of which on its tip cuts off a small oval or reniform spore (sporidium). If the sporidia reach a suitable host plant, they

[63]

produce a short germ tube which bores through the wall of the epidermal cells in order to develop a mycelium in the leaf or bark parenchyma, whose spore bearing structures break forth externally in the shape of aecia and spermogonia. The former are distinguished by the fact that the rows of spores cut off by the basidia (sic) of the hymenial layer are surrounded by a skin-like peridium (*Aecidium Pini*). The aeciospores send their germ tubes into the stomates of their host plants on which they are dependent, in order to produce a new mycelium in the intercellular spaces of these. The fruiting body, which is produced by this mycelium, is called the uredium, and is cushion-shaped; the spores (summer spores, urediospores) are cut off individually or in rows on club-shaped sporophores.

In many rust fungi the urediospores germinate immediately, as the aeciospores, and send their germ tubes likewise into the stomates of the host plants, where, after about 8 days, they develop further uredia. The latter usually lack the peridium characteristic of the aecia. During the summer the urediospores reproduce continuously this form of the fungus, by which means the rust is able to spread very rapidly. In late fall the same mycelium which has produced urediospores produces more teliospores which winter.

The rust fungi with complete polymorphism thus have three forms, namely: the teliospore fruiting structure, the aecium and the uredium. To these belong, for example, *Puccinia graminis* with *Aecidium Berberidis* and *Uredo linearis*.

For some rust fungi, however, it has been shown that besides the teliospores, which apparently are never absent, only the aecia or only the urediospores are produced or that the teliospores produce the same fungus form directly from the sporidia which arise from their promycelium (*Chrysomyxa*).

The rust fungi of conifers, whose number up until now has been rather limited, Rees* divides into four sections which are formed according to the degree of completeness of our knowledge of their alternation of generations.

I. Species with separated alternation of generations (Teliospores on Juniperus species and aeciospores on Pomaceae).
   1. *Gymnosporangium fuscum* Oersted on *Juniperus Sabina* etc. with *Roestelia cancellata* on *Pyrus communis* etc.
   2. *Gymnosporangium clavariaeforme* Oersted on *J. communis* with *Roestelia penicillata* on *Pyrus, Malus, Crataegus* etc.
   3. *Gymnosporangium conicum* Oersted on *J. communis* with *Roestelia cornuta* on *Sorbus aucuparia* etc.
II. Isolated teliospores with direct reproduction.
   4. *Chrysomyxa Abietis* Ung. on *Abies excelsa*
III. Isolated aecia with unknown types of teliospores.
   A. Forms of the group of *Peridermium Fr.* (Bark and needle inhabitors).
   5. *Aecidium elatinum* A & S on *Abies pectinata*
   6. *Aecidium Pini* Pers. on *Pinus sylvestris*
   7. *Aecidium abietinum* A & S on *Abies excelsa*

---

*Die Rostpilzformen der deutschen Coniferen. Halle 1869.

8. *Aecidium columnare* A & S on *Abies pectinata*

9. *Aecidium coruscans* Fr. on *Abies excelsa*

B. Cone inhabiting forms

10. *Aecidium conorum Piceae* Rss. on *Abies excelsa*

11. *Aecidium strobilinum* Rss. on *Abies excelsa*

IV. Isolated urediospore forms with unknown types of teliospores.

12. *Caeoma pinitorquum* A. Br. on *Pinus sylvestris*

13. *Caeoma Abietis pectinatae* Rss. on *Abies pectinata*

14. *Caeoma Laricis* R. Hrtg. on *Larix decidua*

The previous review shows how little we have been familiar until now with the processes of the development of the conifer rust fungi, since for species 5 to 14 we must still find at least one form which produces the teliospores. Probably these are already known fungi whose genetic relationship with previously mentioned fungus forms must only be proven. Only when the developmental processes of the rust fungi are fully elucidated will it be possible to answer the question as to which means are the most suitable for the control of the diseases produced by them.

I have not been able until now to discover the alternation of generation of *Aecidium Pini.* The following reports direct themselves only to the aeciospore form of the parasite.

This occurs not only on the needles of Scots pine (*acicola*), but also on the bark (*corticola*).

Although differing in numerous respects from one another, the two forms are thought to be only varieties of the same species whose differences are conditioned by the type of tissue in which they develop. My opinion whether the needle and bark inhabiting forms are two independent species or not, I will only state somewhat later and provisionally combine the two forms.

THE MYCELIUM

According to my experiences the mycelium of the *Aecidium Pini* grows in the needles of *Pinus sylvestris* and *Laricia austriaca,* as well as in the bark, phloem, and xylem of *Pinus sylvestris* and *P. strobus.*

It consists of hyphae 3.5-4.5 microns in diameter, which in their youthful condition are filled with colorless plasma and show no double layered walls (Fig. 18, 19a, 21c). Older hyphae show clearly double-layered walls, numerous cross walls, frequent branching, and have colorless, only in the proximity of the aecidial fruit body gold-yellow colored drops (Fig. 13, 14, 15, 18, 28, etc.).

Finally, instead of the pure plasma contents, abundant cell sap appears which takes up the greatest part of the lumen of the hypha, while the plasma is forced against the wall and the individual bubbles of cell sap are divided from one another, as a result of which an apparent septation of the hypha occurs (Fig. 21).

The mycelial threads are intercellular, penetrate between the parenchyma cells of the needles, bark, phloem, and medullary rays of the phloem and xylem, between the sieve fibers of the phloem, but not between the xylem fibers (Fig. 19a-22). Sometimes the cells of the parenchyma tissue are completely forced apart from one another by an abundant mycelial development (Fig. 19a). Numerous small branches (haustoria) penetrate into the interior of the parenchyma cells (not the fibers) (Fig. 19, 21, 22h.h.), which bring about the transformation of the cell content, as we will later learn. The mycelium is perennial, under some circumstances making possible reaching an age of 70 or more years. In the needles of Scots pine it reaches at most 2 years of age, because the attacked needles just as those without the parasite die and fall at the end of 3 summers. The mycelium of the bark-inhabiting form spreads radially from the infection point annually in the phloem and cambium until it has encircled the entire stem, which under some circumstances takes a long time.

Differences in the mycelium of the two forms *acicola* and *corticola* are not noticeable.

THE FRUIT BODIES AND REPRODUCTIVE ORGANS

The weather has a great influence on the appearance of the spermogonia and aecia of the needle inhabiting forms, while the time of fructification of the bark blister rust appears to be independent of time.

In 1872 the warm spring stimulated the aecia and spermogonia of the needles already in April, while only at the beginning of June did the aecia appear on the bark.

In 1873 the cool weather delayed the development of the needle rust until the middle of May, and the middle of June still found some aecia on needles not reduced to dust. The form *corticola* on the contrary developed precisely at the same time as in 1872. In general one can say that the form *acicola* fruits during the month of May, the form *corticola* during June.

A statement of Rees: "The aecia appear on the needles in June and July," is therefore to be corrected.

The spermogonia, which precede the development of the aecia by some time, until now were known for the form *acicola* but not closely investigated. They form more or less numerous (up to 50 on a needle) up to 1 millimeter long yellow-brown flecks (Fig. 1b), which are irregularly distributed particularly on the inner side; smaller amounts also on the outside on the 1- and 2-year-old needles. They arise as superficial conical structures on the surfaces of the needle, later break through the epidermis in a narrow longitudinal crack in order to liberate the spermatia. The structure of the spermogonium (Fig. 15) shows no characteristic whereby it can be distinguished from the spermogonia of most other Uredineae. The mycelial hyphae form a pseudoparenchyma between the leaf parenchyma and the epidermis which borders on the thick-walled, phloem-fiber-like subepidermal cell layer (Fig. 15 b.b.), from whose upper surface numerous fine hyphae develop which are inclined to the top of the truncated cone-shaped spermogonium. On the end of these, extremely small 1-2 microns long egg-shaped spermatia are cut off (Fig. 16), which then deposit themselves on the

[66]

two sides of the crack on the outer surface of the leaf.

For the discovery of the spermogonia of the bark-inhabiting form I acknowledge a consignment from Forester Koltermann at Wildenbruch, to whom this opportunity allows me to express my thanks for his repeated interesting reports and consignments. He sent me a twig of a 14-year-old eastern white pine attacked by *Aecidium Pini*, on whose bark between the large aecia were scattered roundish, about pea-sized smooth spots detectable by their somewhat darker coloring (Fig. 7b.b.). These flecks appeared especially frequently on the border between the attacked and healthy portions of the bark. After I had first recognized these flecks as spermogonia, I found them too on the bark of the common Scots pine as to be sure they are quite difficult to detect (Fig. 2b) since the difference in coloration is extremely slight. The spermogonia of *corticola* form a round spot 3-7 millimeters in diameter.

On the border between the bark parenchyma (Fig. 20a) and the periderm layer in the whole expanse of the spermogonia there develops a rather smooth layer about 40 microns high (Fig. 20 c.c.), which towards the interior consists of pseudoparenchyma, similar to that denoted in Fig. 15b, from which fine parallel hyphae arise at right angles to the surface and force the cork layer away from the bark. The latter becomes loose in the vicinity of the spermogonial layer, so that it becomes quite free and the spermatia can develop unhindered, or the entire stroma remains covered by the cork layer.

Here and there one notices in older spermogonia small dot-shaped points of rupture of the cork layer (Fig. 20d) through which secondary fungi have broken forth toward the outside. In the illustration the space taken up by these intruders is shown free. It is doubtful to me, however, whether these points of rupture must be considered as natural openings of the secretion of the spermatia or if they only occur by accident.

The aecia of *acicola* occur between the spermagonia in much smaller number (seldom above 15) on two- to three-year-old needles. Since the attacked needle is not killed by the parasite, thus on three-year-old needles new spermogonia and aecia very often occur scattered between the resin impregnated and dark-brown colored wound spots produced by the previous year's fruit bodies. The crack in the epidermis from which the peridium of the aecium grows forth has a length of 1-3, seldom up to 5, millimeters.

The aecia arise on a loose stroma from which thickly compacted 4-5 celled hyphae (Fig. 13b) or basidia (sic) arise at right angles through the surface of the needle. The outermost layer of cells (Fig. 13p) grows together between one another and forms the first rudiment of the skin-like, later sack-like peridium.

With the enlargement of the aecium it grows in the following manner: in the entire extent of the hymenial layer a row of basidia cut off not the customary spores but bring about from the bottom (Fig. 14p) outward growth of the peridium.

The cells of the peridia are formed similar to the spores, attain a significant size, are polyhedral and are mostly all empty of plasma, therefore transparent. Beyond the basidia which bear the peridium still more club-shaped cells arise against the epidermis (Fig. 14e), and push this towards the outside so that the peridium itself does not have to overcome the pressure of the epidermis. The cutting off of spores from the tip of the basidia (Fig. 14b) does not happen directly one on the other but between each two spores there occur in a youthful condition disjunctor cells of roundish or short or egg-shaped form (Fig. 14d), which disappear after the complete formation of the spore. The sack-like peridium which swells forth through the ruptured epidermis later splits irregularly and permits the loose spore dust to fly out.

The spores (Fig. 17, 18) are oblong, sometimes egg-shaped or polyhedral with truncated angles and corners, which can be explained by the opposing pressures of the rows of spores. The interior is rich in gold-yellow colored plasma, in which swim large and small oil drops (Fig. 17a). The spore wall is colorless, the wart-like thickenings often sit so loose that by a slight pressure of the coverglass they loosen themselves and swim forth freely as small rods somewhat constricted in the middle. Germination occurs very easily within 12-24 hours in moist air as well as in water with one to three thick germ tubes (Fig. 18), which branch numerously, and which often in the plasma still contain the gold-yellow colored oil drops of the spore content.

The aecia of *corticola* are significantly larger and of irregular shape (Fig. 2,3,4,7).

The roundish ones have an average diameter of about 5 millimeters, the longish ones are of about the same width; they fuse together several times with neighboring aecia and obtain a length of up to 15 millimeters. They appear in June on the bark of branches and twigs of young and old Scots pine (Fig. 4), very often, however, also on the stem of the tree itself (Fig. 2,3,5). On younger stems and thinner twigs already in the first year the bark is infected in its entire circumference by the parasite; on thicker branches and on stems of older Scots pine, on the contrary, mostly only one or the other side is attacked, seldom are several sides of the tree attacked simultaneously (Fig. 5). If the tissue of the bark is not completely killed in the attacked spot in the course of the next year, thus new aecia appear between the usually resin impregnated wound spots.

Preferentially, however, these break through in the vicinity of the spot that was diseased or killed in the previous year, as far as the mycelium in the meantime has spread further in the bark tissue.

The formation of aecia repeats itself for a number of years until the attacked plant or branch and so forth has been killed.

On older trees formation of aecia ceases in the diseased spots after a certain time; the mycelium spreads further in the bark; however, it becomes sterile.

Since in the upper part of a tree trunk thick bark formation does not occur even with greater age, on the contrary, the thin bark scales are shed, thus the ending of aecial formation cannot be explained by the thickening of the bark.

The aecia are differentiated from those of *acicola* not only by size and external form but also by the abundantly developed stroma (Fig. 19 a.a.). The basidia which originate from these (Fig. 19b) are significantly thinner, and because they occur thickly pressed together, they develop on a uniformly large hymenial surface with considerably more spores than in the case of the form *acicola* (Fig. 14). The result of this is that in the restricted space the disjunctor cells often appear quite elongated (Fig. 19d), also indeed the spores sometimes are somewhat longer than in the case of *acicola*. If the shape of the lamella was always elongated (Fig. 19), thus under the simultaneous consideration of the other differences the right would arise to set forth the two forms as independent species. In many cases the disjunctor cells, however, are formed just as in the case of *acicola* and I think therefore the question whether *acicola* and *corticola* are true independent species must for the time being remain unanswered until someone has been successful in determining the teliospore forms to which these belong.

Speaking for the difference of the species

1. The size and exterior form of the aecia and the spermogonia.
2. The difference in the size of the basidia.
3. The different time of fructification.
4. The circumstance that in sites where one form occurs in great distribution and frequency the other form very often is found not at all or very rarely.

I lean in general to the opinion that the two forms are independent species for which one may leave the names *Peridermium Pini acicola* and *P. Pini corticola*.

THE MODE OF LIFE OF THE PARASITE

The complete story of the development of *Aecidium Pini* can be given only when we have discovered the fungus forms which belong to it, which, during the time of year between the dissemination of the aeciospores in May and June and the still not established time of the origin of new mycelium of the aecia, perhaps grow on entirely different host plants.

The symptoms of the disease, which are produced by the mycelium by which the aecial form is known to us, are of a quite varied nature, can be explained, however, essentially by the transformation into turpentine of the starch of the cells which the haustoria penetrate.

The form *acicola* (Plate IV, Fig. 1) appears in great distribution in the month of May in young stands of Scots pine.

The most affected are 3-10-year-old young plantations, which often, because of the numerous amount of aecia, have a yellow appearance; scarcely a healthy needle can be detected in the whole stand until in June the healthy appearance of the plants is established.

The mycelium grows so abundantly in the intercellular spaces of the green needle parenchyma (Fig. 13-15) that finally the parenchyma cells become pressed together and compressed without the cells being killed.

In healthy Scots pine needles the chlorophyll-containing parenchyma cells contain no starch, which, on the contrary, is deposited only in the chlorophyll-less parenchyma which surrounds the central vascular bundle. Any significant change of the leaf cell tissue and its contents by the fungus mycelium is not detectable. The cell tissue is killed and completely resin impregnated only in the closest proximity of the spermogonia and the aecia after these turn to dust, so that a dark brown fleck on the needle which itself remains green marks the spot where the fruit body of the parasite has developed.

Even after repeated fructification of the mycelium, the needle which passes into the third year does not die at once, on the contrary, remains still green and living for months, although the mycelium has compressed the green leaf parenchyma so much that much more room is filled up by the first than by the latter. The death of the attacked needles occurs on the average a few months earlier than that of the non-attacked needles, although scarcely any noticeable disadvantage to the health and growth of the host plant can be traced to this.

Here and there one sees indeed the end of an attacked needle or those killed off completely in the course of the second year, yet it remains mostly doubtful whether this dying can be ascribed to the parasite alone, since also unattacked needles often completely or partially die prematurely.

The laminar parenchyma cells of the needle on the upper and under side, which are enclosed for one and a half years by the fungus mycelium, remain completely living, from which must be concluded that there is only an extremely slight effect of the fungus on these.

A killing of the leaf cells would occur when, unlike spruce (Plate VI, Fig. 5), the parenchyma forms continuous cross layers which form a continuous connection from the periphery, the epidermis, towards the interior to the central vascular bundle.

The older the Scots pine, the more rarely one sees aecia on its needles; after twenty or thirty years of age needle rust becomes a rarity.

Aecidium Pini corticola has the most detrimental effect on the health and life of Scots pine. With the description of the parasite its appearance on twigs and branches as well as on the stems of young and old Scots pine has already been discussed.

Stem portions older than 20-25 years in general appear not to be attacked any more and these only in the upper parts of older Scots pine where the bark remains thin by the spontaneous shedding of the scales. In what manner the infection proceeds, obviously at this state of knowledge of the developmental processes of the fungus is an open question. Although, however, very often the nodes (Plate IV, Fig. 3) form the starting point for the disease, this, however, is not always the case; on the contrary, one often sees aecia coming forth from the bark in the middle between two nodes. The mycelium grows intercellularly in the parenchyma of the green bark (Fig. 19, 20a), and in the phloem tissue (Fig. 21a). From this it passes through the rays into the interior of the xylem (Fig. 22a) in which, however, it almost never appears outside of the rays or the resin canals (Fig. 28a). Also the mycelium appears to penetrate into the stem never deeper than about 8-10 centimeters. The hyphae never follow any direction other than that described as through the medullary rays. The effect of the hyphae consists essentially in that short side branches (haustoria) are sent into all parenchyma cells (Fig. 19,20,21,22h), which transform the cell contents, and to be sure at first the starch, into turpentine which precipitates dropwise on all the walls of the organs, until these often are filled up and resin impregnated (Fig. 21,22,23). In Fig. 21 the larger part lying on the right is indicated to be free of turpentine, which, after washing with alcohol, enables the intercellular mycelial threads to be detected more clearly. The resin canals with the surrounding thin-walled, starch-containing cell tissue have been destroyed (Fig. 28); the turpentine, which occurred thereby, as well as that turpentine already stored in the canals, contributed to the resin impregnation of the xylem, which advances to the same extent as the mycelium of the parasite spreads in the phloem. On cross-sections through diseased stem parts (Fig. 8,12), the resin impregnation of the wood can be detected by the darker coloration, fatty nature and characteristic silky luster. In the case of thinner stems (Fig. 8,10), the wood is resin impregnated up to the pith; in thicker cross-sections the interior of the stem remains free of resin; the latter forms only in the periphery of the stem a layer of at most 10 centimeters thickness, since in general the mycelial threads are unable to penetrate deeper.

The great abundance of turpentine, which occurs in the xylem in the diseased spots, cannot be explained only by the transformation of starch in the medullary cells and cells which surround the resin canals; moreover, it must be considered that the turpentine from the stem portion lying above the diseased spots contributes in part to the impregnation. Its own weight causes the turpentine to sink downward in the canals even when these are fully healthy and have not been destroyed by the mycelial threads of the parasite which grow upward in them.

The lack of turpentine above the pitch spots in the so-called resin-top (Kienzopf) speaks for this supposition (Fig. 6).

The complete resin impregnation of the phloem and xylem as well as of the cambial layer results in the stopping of the ability to translocate sap and thereby of the formation of the annual rings as far as the mycelium has penetrated into the phloem.

From the killed, dried out, and split open bark, the turpentine flows freely towards the outside and flows down from the diseased spot of the stem or turns to resin between the bark cracks, whereby the diseased spots take on a whitish color.

Annually the mycelium spreads centrifugally in the bark. Whether the growth of this is limited to definite times of the year is difficult to establish.

At the time of annual ring formation in the vicinity of the diseased spot one finds new mycelial threads in very abundant growth in the new tissue layers which arise by the division of the cambial fibers, whereby it must be concluded that in the first half of the summer the growth occurs especially abundantly, further that the entire appearance of the disease is purely the result of the fungus; that the latter does not establish itself only after the resin impregnation of the plant parts. If the mycelium has developed earlier or later in the entire circumference of the stem or twig and resin impregnation of the xylem and bark has followed, then the plant part lying above the diseased portion dies and, to be sure, as a result of the stopping of the ability to translocate sap. In part the resin impregnation and in part the drying out of the wood contributes to this.

The symptoms of the disease are different according to the age of the attacked plant part and according to the spot where the parasite has settled.

Thinner twigs and branches or young stems show the aecia often in the first year in the entire circumference of the attacked plant part. Death results then already after a few years. In younger Scots pine plantations the Scots pine blister rust often shows itself detrimental by the rapid death of numerous stems. In older stands of Scots pine it often attracts the attention of the alert observer in that individual twigs and branches show a red stained killed foliage. Close investigation will almost always show that *Aecidium Pini* has been the cause of the phenomenon. If the stems of older Scots pines are attacked by the parasite within or below the crown, then very interesting phenomena appear, which are generally known as canker, scab or resin top.

In the first year only a spot 5-10 centimeters in diameter is diseased (Fig. 9-12a). In following years the canker spot enlarges; the width of the annual rings as a result of this increases in thickness on the healthy side of the tree, since the sap, with a downward sinking, nourishes the cambium more abundantly here.

There arises to a certain extent a contest between the growth of the parasite, which seeks to girdle the stem from the infection point, and the growth of the tree which, by the increased growth of the still healthy side, strives to escape the pathogen (Fig. 5, 8-12). Since a general decrease of sap intake to the top of the tree is related to the resin impregnation of the diseased spot and the drying of the conducting xylem, thus the top reduces its growth, the foliage is sparser, the width of the annual rings on the still healthy side of the tree diminishes.

Due to this the spreading mycelium wins the advantage and girdles the stem completely, whereby the drying out of the top follows (Fig. 6). Between the beginning year and the final year of this battle under some circumstances more than 60 years may pass away (Fig. 9, 12). Usually, however, it ends earlier. On the attacked portion of the stem arise characteristic stem shapes (Fig. 8,9,10,11) whose cross-sections are shaped especially noticeably, when, as in general is rare, the parasite has attacked both sides of the tree (Fig. 5,12).

The two canker spots usually touch each other earlier on one side of the stem (Fig. 12b) than on the other (Fig. 12c).

If the diseased spot lies below the crown, then after the drying up of the latter the lower stem dies also; if, on the contrary, vigorous, still abundantly foliaged branches occur below the canker spot, then these remain living for a long time after the dying of the resin top of the tree and frequently the uppermost branch rights itself upwards and seeks to replace the lost top (Fig. 6).

The resin impregnated canker spot is very prized as kindling material (Vogelkien) and therefore the woods offender seeks to acquire the lower portions of the resin top. In the vicinity of villages and cities one sees very often older trees with sawed off tops and can as a rule ascribe these damages to an earlier resin top.

During cutting the woods worker seeks to acquire the resin top and as a rule even careful supervision does not avoid this abuse.

The disease is so wide spread in Scots pine stands that it must be recognized as a great calamity. With alert examination in many stands one will encounter 5-10% of all trees with cankers or resin tops. If one considers that besides these during the course of time through thinning and the culling of dry stems (death cutting) numerous trees attacked by the disease already have been removed, thus it appears that the forester has cause to direct the parasite to his attention. The fact that Scots pine becomes so extraordinarily scattered at higher ages is certainly in a great part to be ascribed to the circumstance that many dominant stems fall as a sacrifice to *Agaricus melleus*, many to *Aecidium Pini*.

Unfortunately it is not possible at the present state of our knowledge to suggest rational recommendations against the disease. Only when we know in what form and on what plants the parasite develops the forms that are still unknown to us can procedures against it, based on scientific knowledge, be suggested.

In the meantime the cutting out of attacked stems will be advisable since, with the removal of the aecia, the development and spread of the teliospore forms still unknown to us also will be hindered or impaired. The complete knowledge of the developmental process of the fungus will bring to hand perhaps controls which are more simple and more easy to carry out.

According to my own observations the pine canker or resin top occurs not only in the school forest at Neustadt-Eberswalde, but also in other pine forests, for example, in the forestry district Königs-Wüsterhausen, further at Swinemuende in the dune stands on best and poorest soil classes. In many stands one seeks in vain; in others it is quite frequent, as has already been discussed earlier. Any dependence of the disease on influences of weather and soil is unknown to me. In very dry years more frequent dying of the tops or branches of diseased trees occurs, which is to be considered as a natural result that, after the resin impregnation of the xylem in the canker spots has occurred to a certain degree in the course of the year, sap uptake to the plant parts lying above this is very diminished.

Excessively increased evaporation in hot dry summers must hasten the complete drying of the top and so forth which in wet years still kept itself green.

From the forestry literature I can record here some reports about the distribution of the disease without having to correct in particular the in-part erroneous observations about the character of the disease.

District Forester Wachtel at Neuhaus in southern Bohemia sent diseased Scots pine to the Geheimrat Ratzeburg. It appears that the diseased trees occur there abundantly not only in sapling-sized stands and pole-sized stands, but also in mature stands. It is known by the common folk as "Kozor" and by the Germans "Schorbel," besides at one time as canker or rust, at another time scab.

The diseased trees according to Wachtel were occurring on the most varied soil classes. In 1855-56 Mr. Wachtel noticed dry treetops everywhere in his pine forest; the stems were, however, quickly felled and worked up and so the pest was brought to silence. In 1861 this drying of the treetops occurred suddenly again - apparently favored by the dry year of 1857-59 - and made a careful inspection of the stands necessary.

Ratzeburg found by chance in some of the submitted cankers larvae of a moth which was feeding on the killed bark. He named this *Tinea sylvestrella* and in the superficial and unscientific manner characteristic of him as far as botanical questions were concerned, ascribed the entire phenomenon to the frass of this moth just as he, in a similar manner, ascribed the silver fir rust, which is produced by *Peridermium elatinum*, to *Sesia cephiformis* which was living in the killed bark. The samples in which Ratzeburg found no larvae or larval tunnels made little concern to him; "I have decided in favor of sylvestrella; because the diseased spots show scab, mange, and resin pustules as with the true sylvestrella..."*

The aecia of *Peridermium Pini*, to be sure, also attracted Ratzeburg's attention and he even described through his draftsman (Fig. 18) their symptoms, to be sure, indeed, only as a result of a mania to illustrate all, even when it was completely unintelligible to him.

He said about it: "Finally, however, the small parasitic fungi (Cryptophyten), which are at once conspicuous on living stems due to their orange color, are still to be remembered."

The disease was labeled by Ratzeburg with the name "Mottendürre." Wachtel considers this disease infectious because it spreads if only a few stems are overlooked: if one believes his pole-sized stands and sapling-sized stands to have been well cleaned up, in a few weeks attacked stems always appear again. However, the insect, as Wachtel observed, is lazy in that it always seeks new victims only in the vicinity. This year (June 1863) the "Scots pine fungus" appears frequently in all suspected places, and it will also yield sufficient *sylvestrella* and *piniphilus* (W.). To that Ratzeburg noted: "This red fungus is a symptom well to be considered and I have not omitted it, therefore, in the description of a magnificent example of the "Mottendürre," on which it flourishes." (cf. Ratzeburg, Waldverderbniss, Vol. 1, pp. 196-200).

In pleasing opposition to Ratzeburg's description and explanation of the disease is a work published in Volume 2 of Burckhards: "Aus dem Walde" by the district forester Wissmann of Bovenden near Göttingen. Although the author could not explain the appearance of the disease by the spread and effect of the mycelium, nevertheless, he gives an excellent description of the external symptoms detectable with the unaided eye and recognizes *Peridermium Pini* as the cause of the disease. Numerous inaccuracies and errors are not the way to capture the description of the character of a worthwhile scientific publication. I extract the following as an abstract of the work: The Scots pine canker was observed abundantly in 1855 and 1856 also in Hannover and to be sure on some sites to a threatening extent...The observations made here afforded proof that as to the cause of the canker there is, however, a different state of affairs than Ratzeburg gives, who ascribes this

---

*Ratzeburg; Waldverderbniss, Vol. 1, pp. 196-200.

to *Tinea sylvestrella*; that it simply is a result of the stem being attacked by a rust fungus, *Caeoma Pini* etc., etc.

As a rule the entire stem above the point attacked by the parasite is killed in a space of a year (??).

Only rarely does one see attacked stems on which the rust girdle has not continued completely around, so that in the undiseased spot the circulation of the sap persists between the parts above and below the girdle and life of the stem is still delayed...One notices the Scots pine canker quite easily in young stands; however, it also occurs in the infested localities in the crown of older trees. That it seeks out here only the upper regions appears to be explained by the fact that the hardened and in the outer layers already killed bark, must hinder the penetration of the germ tubes. My own observations and those sent to me by others confirm that preferentially only those stands which occur on very poor and dry types of soil are attacked by the canker. The canker has been not noticed by us yet in well growing, fully stocked stands on moister soil and the well stocked better parts of attacked stands remain free of it. It appears that the healthy pine stem is not prone to the attack of the fungus. Dry hot years, such as that of 1865, seem to cause an increase, as can be gathered by the fact that all at once in the spring of 1866 from so many sites the serious appearance of the fungus and its results have been observed simultaneously...Forestry supervisor Ebeling in Miele has observed the spread of the canker in a northwest direction and has explained this by the fact that at the time of ripeness of sporidia in May southeast winds predominate. Each canker-diseased pine is not only a center for the spread of the fungus but also the breeding spot for insect enemies of Scots pine.

The attacked stems therefore are to be eliminated immediately....

A cutting should be carried out especially before the complete development of the fungus; however, also can occur at any time of the year....

The hope expressed in the report of May 1869 that the careful removal of the stems attacked by the canker (undertaken in the winter of 1866/67?) in the Scots pine stands there has had the result hoped for, since, in spite of the dry summer in 1868, only with effort can any samples be found, appears to me to be perhaps somewhat premature. A cutting out of all diseased Scots pine affects those trees which, in part, have been already diseased for decades; if also in 1867 and 1868 the number of annual fellings due to disease has been equally high as in the previous years, thus an inspection after two years will still be able to encounter only a few diseased trees and these in a weakly diseased condition.

A third publication about the canker disease of Scots pine occurs in "Grunerts forstlichen Blättern", new series, second volume, April 1873, page 113, whereby the author, in a consideration of several disease phenomena of forest trees, published the following:

"On the contrary, an appearance of disease in Scots pine is of peculiar nature and we will direct our attention to this. This has been encountered until now only in a single Scots pine site of the Erlenbach forest district belonging to the Konigl. Oberförsterei Trier."

"The site is perhaps 12 hectares large, forms a northward moderate slope and has good loamy sandy soil, arising from variegated sandstone, of moderate thickness. The Scots pine have originated from seed, now are about 40 years old, growing rather rapidly, indeed, for local conditions, slender and rather dense."

"On a rather large number of these Scots pine stems, scattered through the whole site, perhaps at about 2/3rds of their height and easily detectable by a dry blackish coloration and resin flow, there were 0.5 to 1.0 meter long spots on one side of the stem, at different cardinal points, here and there also a spiral twisting of the stem, and in the course of several years the dying of the diseased stems had always resulted. The dead stems on the site were removed several times during thinnings; nevertheless, this site today contains a moderate quantity of stems which show in different degrees the formation of these wounds and face their death. In the first stage of the disease the diseased spot appears with definite boundaries, somewhat sunken and with curled bark, until resin flows out on it; the spot becomes increasingly dark colored, and the wood layers under it appear as dead. On the side which lies opposite the wound spot new annual rings are put on continuously up until the death of the tree; they thicken themselves toward the outermost point of this continuously until the circulation of the sap at the open spot ceases and the life of the stem expires. The progress of the symptoms of the disease is, if one follows the process of this on a cross-section, always a long one, perhaps taking 10 to 15 years; however, death results only when the wound first becomes noticeable to foresters by becoming black and by the accumulation of resin, as noted in a few years. There is seldom more than one wound on an individual stem, although we have here and there noted two on different parts of the stem of the same tree."

"The disease cannot be traced back to a definite external cause; namely one does not know that the locality at any time has suffered damages by insects or the influences of weather or indeed whether a fungus disease was indigenous in the pine plantations there and perhaps has attacked the locality in question. Investigations of the phenomenon already conducted by competent naturalists will hopefully give details of this relationship."

The author of this article in January of this year sent me diseased pieces of stems from the stand in question which in part have been illustrated (Plate IV, Fig. 5,8,10). On all of these the remains of the aecia still are detectable at the first glance. At the end of July of this year I obtained from the same place a sample of small stems and branches from an approximately 18 to 20 year old stand near a 40 to 45 year old stand which had been diseased for a long time, which had already been killed in part by the Scots pine blister rust.

Explanations of Illustrations

(Plate IV)

Fig. 1. Scots pine needle with aecia (a) and spermogonia (b) of *Peridermium Pini* v. *acicola*.

[76]

Fig. 2. Scots pine branch with aecia (a) and spermogonia (b) of *Peridermium Pini* v. *corticola*.

Fig. 3. Stem of a young Scots pine with a whorl of branches, which have been attacked for several years by *P. Pini*.

Fig. 4. Branch from the crown of an old Scots pine with twigs already killed by *P. Pini* (s).

Fig. 5. Piece of a stem of an about 40-year-old Scots pine with two cankers lying on opposite sides. The killed bark bears the obvious traces of the aecia of *P. Pini* (a.a.).

Fig. 6. An about 80-year-old Scots pine with a resin top affected with blister rust and with a substitute tree top.

Fig. 7. Bark of a branch of *Pinus Strobus*. The aecia in part still not ruptured (a.a.) and partly ruptured (c). The spermogonia (b) detectable as smooth dark flecks the size of peas.

Fig. 8. Cross-section through the canker of a 50-year-old Scots pine. At the time of infection the attacked part of the stem was 25-years-old. For 5 years the infected and thereby resin impregnated spot has enlarged annually so that only the light part is living and conducting sap. Each annual ring is drawn only for the last 5 years, previously each 5 rings are shown collectively.

Fig. 9. Cross-section through a canker of an 110-year-old Scots pine. The attacked stem portion was 15 years old, was infected at (a) the mycelium spread itself in all directions in the phloem, whereby the side of the tree capable of growth became always more restricted. The tree was killed shortly before felling, after it had preserved the ability to translocate sap in the diseased portion of the tree for 70 years by a one sided growth in thickness. The non-resin impregnated area at b was the only area still fungus free. Bark formation occurred only in this spot and on its margin was rather thick; in the remaining area it was very thin.

Fig. 10. Cross-section through a canker of an approximately 50-year-old Scots pine stem. The attacked portion of the stem was 20 years old. During the 20 years of the disease the stem except the two light colored parts (b.b.) was filled with fungus. The presence of a still green branch above the cross-section explains the spot spared by the mycelium until now at (b) on the left side.

Fig. 11. Cross-section below a canker through an approximately 90-year-old stem. The mycelium of the parasite has grown only up to (a) in this part of the stem and has impregnated the wood with resin. The enlargement of the canker lying above, however, has caused the gradual stopping of growth of the parts of the tree lying immediately below so that finally only at (c) does complete growth still occur, above which side on the higher lying canker spot only the bark is still healthy.

Fig. 12. Cross-section through a double canker on an about 110-year-old Scots pine. At about 15 years of age this portion of the stem was attacked at (a) and (a), the mycelium grew to point (b) after 25 years, after 70 years the small spot (c) still is fungus free and not resin impregnated.

Fig. 13. A youthful condition of an aecium on a Scots pine needle in cross-section. The intercellular mycelial threads are septate, branched, next to the aecium with yellow oil drops, form here a loose stroma from which arise the basidia (b) consisting of 4 to 5 cells. The outermost cells of the basidia (p.p.) pressing against the epidermis and finally splitting this open, grow together with one another to form the first layer of the peridium, which later enlarges itself so that in the periphery of the sporophores a row of basidia brings about the regrowth of the latter.

Fig. 14. Part of an aecium of *P. Pini acicola* at full development. The basidia (b) cut off at their ends alternately spores and small membrane layers (d), the latter, with the formation of the spore wall, disappear. Outside of the peridium (p), which consists of hyaline polygonal cells which are

[77]

formed similarly to the spores, one still sees some of the basidia (e) which with their club-shaped end cells press the epidermis towards the outside.

Fig. 15. A developed spermogonium in a Scots pine needle. The mycelial threads form a thin layer of pseudoparenchyma (b) between the leaf parenchyma and epidermis, from which arise slender hyphae which cut off spermatia on their tips and form a flat cone-shaped fungus body.

Fig. 16. Spermatia, which arise by abstriction from the tips of the hyphae of the spermogonium.

Fig. 17. Spores, whose epispore bears short rod-shaped warty thickenings. The latter with light pressure often completely fall off and then the contents of these, consisting of yellowish plasma and large yellow oil drops, are clearly detectable.

Fig. 18. Germinated spores with one or more germ tubes, into which flow the in part still yellow-colored spore contents.

Fig. 19. An aecium of *P. Pini corticola*. The stroma (a.a.) is strongly developed between the bark cells forced apart from one another, in which individual haustoria (h) are growing. The basidia (b) are somewhat smaller celled than in Fig. 14; as a result of this a much larger number of these occur on an equal sized hymenial surface. The membrane layers (d) often attain a stretched-out shape, whereby more room is attained for the development of the spores. Outside of the peridium (p) I have noticed no basidia.

Fig. 20. Cross-section of a spermogonium of *P. Pini corticola* from the bark of *Pinus Strobus*. Between the periderm layer and the outermost layer of bark parenchyma, the intercellular mycelium of the bark (a.a.) develops a very thin pseudoparenchyma layer (d) from which parallel hyphae (c) arise at right angles to the surface, which press against the cork layer and the slender ends bow under these so that the regularly parallel positioning of these disappears. Sometimes one sees in older spermogonia small breaks in the cork layer filled with extraneous fungi penetrating through these. The latter are not illustrated; the space taken up by them remains empty (d).

Fig. 21. Radial section through the phloem of a Scots pine attacked by *P. Pini corticola*. The part lying towards the right (y) has been washed with alcohol, in order to show the intercellular mycelial threads (a.a.) and the haustoria (h) in the parenchyma cells. The content of the latter has almost completely disappeared with the exception of the cell nuclei. In the part marked (X) one sees the abundant drops of turpentine (s.s.) condensing on the walls of all organs.

Fig. 22. Radial section through the wood of a canker. The mycelial threads (a) penetrate through the medullary rays into the interior sending numerous haustoria (h) into the medullary ray cells, passing, as Fig. 23 shows in cross-section, from the medullary rays into the resin canals and destroying the thin walled tissue surrounding these and its contents. The turpentine arising from the transformation of the starch, as well as that existing in the resin canals of the tree portions lying above the canker and the volatile oil precipitating towards the bottom, condenses onto the walls of the wood fibers first in small drops (s.s.) which later flow together and completely fill the interior or leave only unfilled small air bubbles (tt).

Caeoma pinitorquum A. Br.

(Plate V, Fig. 1-9)

The Pine Twister

The disease which is caused by *Caeoma pinitorquum* has in recent years become so noticeable in its distribution and intensity, that I already considered it necessary to publish a paper on this disease accompanied by an illustrative plate in the Zeitschrift für das Forst- und Jagdwesen von Dankelmann 4(1), 1871. As I suppose this work has been seen by only a few botanists, and on the other hand continued investigation has widened my knowledge of this interesting parasite, thus I give here a new treatment of it. What de Bary published in the Monatsberichten der Königl. Akademie der Wissenschaften zu Berlin in December, 1863, about this new parasitic fungus of Scots pine sent to him by A. Braun, bears the character of a hasty investigation of poor and especially too-old material. Now if de Bary was not self-evidently responsible for the condition of his investigated material, this explains on the one hand the incompleteness of his investigations and on the other hand the creeping in of numerous errors.

He did not observe the spermogonia which are very striking, and also the entire process of spore formation was conceived incorrectly. Unfortunately, I have not been successful in establishing clearly the process of development of this rust fungus in its different forms and I refer to what has been said about the rusts in general on page 63 and forward. Accordingly, *Caeoma pinitorquum* at least still lacks a teliospore form, perhaps possesses this but also still lacks a third form, the aecial form, since a *Caeoma* can be interpreted as a *Uredo*, whether correctly may be left undecided for a while.

THE MYCELIUM

The mycelium grows intercellularly, preferably in the green bark parenchyma of young Scots pine shoots, from which it spreads outwards into the phloem and through the medullary rays to the pith of the young shoot.

With young seedling plants (Fig. 1) it develops also in the leaf parenchyma of the cotyledons.

The hyphae (Fig 6, 8) have a diameter of about 2.5 to 3 microns; when young they are filled with a colorless plasma, and after the disappearance of this show a clearly detectable thickening of the cell wall; in that case the septa are detected only with difficulty (Fig. 8e). In the vicinity of the fruiting body and spermogonium the contents of the hyphae are colored yellow gold; as a result of this, the portion of the green bark on which the fungus prepares itself for the formation of the fruiting bodies by abundant development of the mycelium is detectable externally very early by the yellow coloration. The mycelium sends numerous short branches (haustoria) into the interior of the parenchyma cells (Fig. 6h), moreover, is clearly branched in the intercellular spaces.

The life span of the mycelium generally is very brief, since after sporulation that mycelium under the fruiting body dies with the cell tissue up to the pith. Indirect proof exists with great certainty, however, for the opinion that a portion of the mycelium is perennial in the twig and grows forth yearly into the newly developing shoots. In which tissue parts the perennial survival and growing forth of the mycelial threads occurs is at this time still an open question.

FRUIT BODIES AND REPRODUCTIVE ORGANS

According to the weather in the second half of May or the beginning of June, when the new shoots of Scots pine have developed to the point that the green bark of the stem appears between the needle sheaths in the lower portion, the needles themselves seen with their tips scarcely out of the sheath, one detects spots on the bark, originally whitish, later becoming yellow, the future fruiting bodies in their earliest stage of development. Microscopic investigation shows in most cases spermogonia in a quite advanced condition, the uredio body, on the contrary, often still absent, often in the first stage of development (Fig. 5). (His use of the term uredio must be an error, since he obviously must mean the aecia WM.)

The spermogonia (Fig. 5, 6, 7sp), which are distributed in great numbers on the discolored portion, arise in the following manner: the mycelial threads penetrate not only between the bark parenchyma cells but also between the cells of the epidermis, push these apart, and between the cuticle (Fig. 6c) and the epidermal cells (Fig. 6e) form a broad cone-shaped fruiting body which consists of numerous fine hyphae undulating and bending to and fro and tending towards the point of the cone. The cuticle is pressed somewhat towards the outside by the spermogonium so that even at low magnification the small point-shaped organs are detectable on the epidermis of the bark (Fig. 2a); the epidermal cells, on the other hand, are pressed mostly somewhat toward the interior, however, sometimes disappear almost entirely in the thick tissue of the spermogonium. The spermatia, which are enlarged more in Fig. 6s, are liberated from the cuticle which splits open at the tip of the cone.

Below the yellowish translucent epidermal spot filled by numerous spermogonia, one sees in the second or third row of parenchyma cells the first rudiment of the urediospore fruiting body forming (Fig. 5). This reaches a length of one to two centimeters and differing widths. Sometimes it appears as a narrow, gold-yellow stripe, often as a broad spot engaging a fourth of the twig circumference (Fig. 2aa). Not uncommonly so many fruiting bodies arise together so thickly that they appear as a very broad body.

The mycelial threads penetrate in great numbers from the interior between the cells of the parenchyma cell layers, so that these are surrounded on all sides by hyphae (Fig. 6st), which end below the next row of cells lying further outward, and are thickened a little on the end. In Fig. 5 & 6 the third parenchyma cell layer is the one in which the basidial layer forms, so that a two-cell-wide bark layer covers the new fruiting body. (He is using the term basidium to mean sporogenous cells in the base of the aecium WM.) In Fig. 7, in which a further stage of development is illustrated, on the contrary the second bark cell layer was the place of origin of the fruit body. The cells, which are surrounded by the basidia, later appear in part to be entirely reabsorbed; at times one still detects between the basidia, which are arranged in clusters, small holes which are interpreted to be the residue of these cells (Fig. 8ff).

The characteristic arrangement of the rows of spores, not completely parallel but converging towards the inside or the bottom (Fig. 7, 8), is due to the origin of the basidia in the intercellular spaces of the bark cells.

Spores are abstricted on the tips of the club-shaped basidia (Fig. 6, 8b); they are very thin walled in the beginning and are separated from each other by flat disjunctor cells (Fig. 8c). With the enlargement of the spore and the complete formation of its cell wall, as in the case of related species here also the intercalary cells disappear (Fig. 8d). Whether the immature spores are surrounded by a hyaline jelly layer as de Bary reports I must leave undecided, since I have never been able to detect such. Since the entire illustration of de Bary's page 627 of the mentioned volume must be pointed out as one which is incorrect in the essential points, the existence of the intercalary cells was not known by him, etc., so also a doubt as to the correctness of the above allegation must be justified. The abstriction of the spores is successive; while the first formed spores are already completely formed, continuous abstriction of spores and disjunctor cells still occurs on the ends of the basidia (Fig. 7, 8).

With the formation and enlargement of the spores, which may arise at least twenty in a row, the fruiting body increases in size; the uppermost spores force the covering bark and epidermal cells toward the outside and thus a welt-like swelling of the yellow spots occurs. The pressure that the growing fruiting body exerts acts simultaneously towards the interior of the deeper lying bark tissues, which are also pressed together thereby toward the middle of the twig, by which then a hole filled with spores arises in the bark. In thin cross-sections, the upper loose spores fall out and thereby an empty cavity is formed between the covering bark and epidermal layers and the spore body. Finally, indeed, at the beginning or towards the middle of June, the fruiting bodies split open in a longitudinal crack (Fig. 2a) and the spores are scattered outwards.

The spores are mostly spherical, sometimes oval or rounded off, polygonal, 15-20 microns large. The contents are fine grained, pale yellow-reddish; the wall is colorless and consists of two layers. The inner layer is homogenous and forms a light zone below the outer layer which appears as if it was joined together from numerous radially positioned equally sized little sticks. These, however, do not, like the spores of *Peridermium Pini*, loosen with the application of pressure from the layer below, moreover seem to be completely joined with one another. After I had failed many times to bring the spores to germinate artificially, I found germinating spores (Fig. 9) sticking externally in the immediate vicinity of a uredio fruiting body which had been ruptured open for a long time on the dead bark surface. The germination showed no differences from other urediospores.

After spore formation ceases, the basidia still lengthen themselves significantly into long colorless, club-shaped tubes. After the rupturing of the fruiting body, the epidermis and bark layer above it dry up, and roll together towards the margin, or are entirely cast off, especially if several fruiting bodies arise close to one another. At the end of June, as a rule, the complete development of the fungus, as we have described it until now, has ended. Where and in what form our parasite occurs from the beginning of July to the end of May of the next year we still do not know, although later I will come back to this question again.

Soon after the dissemination of the spores, the cell tissue occurring in the immediate vicinity of the fruiting body dies, becomes brown colored, dries up, or is pitch impregnated. How far the dying of the shoot extends in one or another direction from the boundary of the actual fruiting body depends upon the distribution which the mycelium has attained in the tissues of the bark, phloem, xylem, and the pith during the development of the fungus.

As a rule, the outer part of the green bark dies for several millimeters on the periphery of the earlier yellow fungus spot, becomes brown and shrinks together. On several year-old fungus wounds already walled-over, this killed tissue still covers the unkilled portions of the green bark (Fig. 4a). The assertion of de Bary that to the extent that the outer surface becomes brown, so also the underlying tissues of the twig and, indeed, the bark, cambium, xylem and pith die, is true only in rare cases. Extruded resin often completely fills up the hole of the fruiting body and numerous rot inhabiting fungus forms develop on the killed tissue.

Except for in the case of quite small, insignificantly developed fruit bodies, towards the interior the bark, phloem, xylem and pith below the fruit body die and become colored brown (Fig. 4x).

In the pith the brown coloration often extends 1-2 centimeters above and below the fruit body, a proof that the fungus mycelium has developed extensively and vigorously here.

THE MODE OF LIFE OF *CAEOMA PINITORQUUM*

I have observed *Caeoma pinitorquum* until now only on the common Scots pine and, to be sure, already on very young, few-week-old seedlings (Fig. 1), on which not only the subcotyledonous stem (Fig. 1a) but also the cotyledons

(Fig. 1b) and the small primary needles were attacked. In a large pine seedbed of the Count Schulenburg Forest at Neustadt-Eberswalde, I discovered on June 29, 1871, that of the vigorously germinated little plants at least two thirds were attacked by *Caeoma* and the uredio fruiting bodies were already releasing spores. This indicates the noteworthy fact that the spores through whose germination *Caeoma* arises must be disemminated at the beginning of June or towards the end of May, since the young plants make their first appearance only after the middle of May. Those attacked plants on which the fungus had developed only on the cotyledons, grew undamaged and vigorously after the drying up of these leaves; those plants attacked on the stem, for the main part, perished; a proof that the fungus is not the symptom of an internal disease but the cause of a local dying. A 10-year-old Scots pine plantation immediately adjacent to the seedbed and interspersed with aspen and birch has already been severely diseased for four years, whereby it must be concluded that conditions for the development of the parasite were present, that is to say, the plants on which the teliospore form develops.

An interesting opinion on the question of when the infection on the plant occurs was sent to me by forester candidate Roloff. I extract the following passage from his letter: "It can be proven here (forest district Ziegelroda), that this year (1872) infection by spores must have occurred not at the beginning of June but in the period from April 25 to May 18. From a nursery of two-year-old Scots pine, a planting was made not only in the nearby bordering district but also plants from the same nursery were utilized somewhat removed in another district on April 25 of this year. The latter are, as I have convinced myself, completely healthy, while those remaining in the nursery as well as those planted in the neighborhood were attacked by *Caeoma*.

On the contrary, this years sowing of Scots pine in this nursery, which began to come up on May 18, is in no way infected."...

This indicates that the infection could not have occurred before April 25, otherwise the plants dug up and sent away shortly before April 25 would have been diseased as were those Scots pine remaining in the nursery or planted out in the vicinity; further, no general diseased condition of the plants has produced the fungus; on the contrary, this must have come as a result of the landing of spores.

The very warm spring weather in 1872 has given the vegetation an advance of 8 to 14 days, so that the infection, usually occurring at the end of May, indeed, can already be ended by the middle of May and the seed which germinates later remains spared.

One-year-old and older Scots pine are regularly attacked only on the young shoots (Fig. 2), never on the needles. The younger the plant, the more dangerous the parasite is to it. Most frequently 1-10-year-old plantations are attacked; rarely does the disease occur initially in 10-30 year stands, never in older stands.

From the previous description of the fungus, its external appearance is known on the young shoots occurring in May. Where the fungus occurs in a Scots pine plantation for the first time, it never occurs there in large amounts but only a certain percentage of the plants are attacked; on these only individual shoots are diseased, on these shoots the gold-yellow flecks occur only sporadically.

The twig almost never dies as a result of the scattered infected spots but at the diseased spot there occurs a bending which gives the twig an S-shape when the upper part grows upward again (Fig. 3aa).

This characteristic has given rise to the choice of the name "Kieferndreher" (Scots pine branch twister).

This wound is usually already healed over after one year (Fig. 4x), on the healed-over swelling, with very vigorous growth of the shoot, after 4 to 5 years it can scarcely be detected; with less growth, however, especially on side branches of older Scots pine, it is still detectable after 10-12 years. It is distinguished from healed-over wounds of weevil feeding injury, which are easily confused with it, by the fact that on cross-sections the pith, xylem, and phloem of one-year-old shoots are colored brown under the wound spot, while the feeding points of weevils almost always arise in shoots of greater ages and no dark brown coloration is produced in the xylem lying further towards the interior up to the pith.

One- and two-year-old Scots pine still possess such thin shoots that only one fruit body is sufficient to kill their entire circumference.

If, however, the young plant is still living due to the development of buds in the leaf axils or the development of sheath buds, the repetition of the disease phenomenon, which occurs almost without exception and in increased incidence during the following years, is so injurious that diseased young plantations are almost completely lost. If, on the contrary, older plantations are attacked by the parasite, the disease even in three-year-old plants in the first year usually shows no very virulent character, since, as already mentioned, in the beginning only individual shoots bear sporadic fungus flecks. Often in the second year, only first after a longer period of time, the disease can take on an intensity so that not only all plants but also all shoots on the lower and middle part of these are so covered by the fruiting bodies of the fungus that usually the shoots completely die with the exception of a short stump (Fig. 3d, 4). Often one can observe on one twig all degrees of the damaging influences of the parasite. Either only crookedness of the shoots occurs, which shoots, however, recover by the fact that buds form and develop new healthy shoots in the next year (Fig. 3aa), or the twig is so weakened that the buds sprout only scantily (Fig. 3b) or die entirely (Fig. 3c). If the shoots are infected on all sides with numerous fungus spots, they also die completely (Fig. 3d, 4a) or only the lower most part of these remains living (Fig. 4b). In this case in the next year several of the lateral buds borne on the short shoots between the needles develop into lateral shoots (Fig. 4b) or dormant node buds develop (Fig. 4c).

Plantations in which the disease is occurring quite intensively give the impression in July that a late frost has killed the new shoots. Stands diseased for several years give the picture of a place severely browsed on annually by red deer. Plantations which were attacked before 6 to 8 years of age in most cases can be considered as lost, since they become completely deformed. Plants attacked at later ages show frequent bending of the branches or stems; however, they are not diseased to such a high degree as young Scots pine.

The danger of the disease lies above all in the fact that the plant, once attacked, becomes diseased again annually. The intensity is, however, quite different from year to year and depends entirely on weather conditions. If the spring is wet and cold, then the parasite develops in an uncommonly abundant manner. The fruiting bodies appear in great number and in marked perfection and size so that frequent dying of the twigs results. If the spring weather is warm and dry, then the development of the fungus is very much retarded, indeed, the formation of the fruiting bodies in most cases does not progress beyond the first rudiments. In this year the disease is so insignificant even in the stands most attacked that it is scarcely noticeable. With more exact investigation, to be sure the rudiments of the fungus spores are detectable on the new shoots by the yellow coloration; in many cases their formation scarcely came up to the stage of development illustrated in Fig. 5. The spots later became shriveled; spores did not develop at all. This suggests the thought to explain this by the fact that the fungus mycelium can develop abundantly only in the tissues of the plants when the transpiration of the Scots pine, hindered to a great degree by the cold wet weather, has brought about a higher saturation of the cell tissues of the plant with moisture, whereas a vigorous transpiration in dry air brings about a deficit of moisture unfavorable for the development of the fungus. It is probable that by a series of dry years the development of the fungus can be so hindered that the once diseased plants recover; further, that the less intensive occurrence of the disease on older Scots pine is related to a lesser abundance of sap in the young shoots; however, these are only opinions which I would like to put forth to explain a so far otherwise unexplained phenomenon.

I conclude from the type of disease symptoms that the fungus is perennial in the interiors of a plant once attacked, from the previous years shoot of which the mycelium grows forth into the buds or new shoots and here is able to develop fruit bodies without new infection from the outside. If a Scots pine, which has been healthy until now, is attacked by the disease for the first time, it shows fungus spots only on some twigs and to be sure almost always only on that side of the shoot which is turned towards the center of the disease (more about this later). Already in the second year almost all node shoots and middle shoots from the twigs diseased in the previous year show the disease, which is no longer limited to one side of this branch.

The fungus spots now are distributed on all sides in rather equal amounts; they are also very numerous and occur each year from anew, even if, indeed, at different intensities according to weather conditions. I know of some stands in which it can be demonstrated on the still recognizable wound spots that for 12 years the disease has returned annually without interruption.

The often very definitely detectable annual progress of the disease is very noteworthy as it is especially suitable to disprove any eventual trials to derive this from influences of soil or weather. One example may be mentioned here.

[85]

Near Neustadt there occurs a Scots pine stand already about 20 years old of about 50 hectares, which is surrounded on the west, south and east by older stands; on the north it is bounded directly by a field. Close to the field on the northwest side the disease occurred for the first time 14 years before; up to 1866, the disease spread very slowly over an area of about 5 hectares size; from 1867 onwards the disease rapidly spread toward the south:

1867 about 50 paces
1868 about 70 paces
1869 about 50 paces
1870 about 130 paces
1871 about 160 paces

which means up to the south border of the stand, which now is diseased all the way through. There are known to me nearly 40 stands, which are more or less severely diseased. The particular observational results, published on pages 110 to 122 of the cited issue of the Zeitschrift fur Forst- und Jagdwesen, I will not give here again, but repeat only those which seem to be noteworthy as an indication of the still-to-be-proven relationship with a teliospore form, namely the fact that without exception all diseased stands examined by me lie directly beside or so near to a field that infection can result easily from a fungus growing on field plants or field weeds. Where the disease occurred first in a plantation it was always on the field side from where the parasite penetrated deeper into the stand. In the first year almost without exception the diseased spots occurred on the side of the shoot which was oriented toward the field; on the outermost boundary of the actual zone of spread away from the field, only the most vigorous Scots pines projecting above the average height of the plantation lying before them were diseased in their terminal shoots, while the lower shoots as well as the lower plants were healthy, that is to say, they were protected from the deposition of the spores. Some exceptions to this rule, in which diseased stands lay in the middle of the forest, confirmed the rule every time in that either small field crop areas occurred in the forest in the direct vicinity or land appropriated for the use of forest officers lay situated to the diseased stands in such a way that under consideration of the prevailing wind direction an infection could very well have occurred from there.

A second fact is still to be mentioned, the almost without exception occurrence of aspen (*Populus tremula*) in the stands which are diseased. The very widely distributed *Melampsora populi* parasitic on aspen leaves apparently cannot come into question, since this is known to belong to the urediospore form, *Epitea populi*, further because poplar rust is also commonly distributed in stands in which *Caeoma* does not occur.

As for the rest, from the observations reported by numerous others and by me are just the facts that the disease occurs here and there throughout all of North Germany, that it occurs on the best and poorest soils, on wet and dry soils, that cold wet weather increases the disease, that dry hot weather retards it. Apart from the quite generally recognized and well explainable stimulation

[86]

of fungus germination and development by wet weather, I have already spoken of the particular influence of this on the development of *Caeoma*.

At the time being I am not in the position to suggest means of prevention or eradication, until I have been able to name the plants on which the spores of *Caeoma* germinate and which produce other fungus forms from July to May.

Explanation of Illustrations

(Plate V, Fig. 1-9)

Fig. 1. Scots pine seedling attacked on the stem (a) and on the cotyledons (bb) by *Caeoma pinitorquum*.

Fig. 2. Young Scots pine shoot (middle of June) with the fruit bodies of *Caeoma* at (aa). The needle fasicles which protrude from the sheath have already been plucked out with the exception of the uppermost ones. On the thick, yellow-colored bark spots, before bursting open, the somewhat raised point-like spermogonia are detectable.

Fig. 3. Twig of a Scots pine diseased for 15 years. The fruit bodies have resulted partly only in a crooking of the shoots at (aa), partly a weakening of the next year's shoot at (b), or even a killing of the buds at (c), and finally completely killed the shoot at (d).

Fig. 4. Twig of the same tree, whose previous years' shoots have been killed either totally at (a) or in the upper part at (b). On the living lower portion of this the lateral buds have developed into lateral shoots at (b), at the node dormant node buds have developed into new node shoots as at (c).

Fig. 4x. Cross-section through a nearly walled-over wound spot on a two-year-old Scots pine. At (a) the bark which has been killed in the vicinity of the fruit body after sporulation of the fungus under which the bark, phloem, xylem, and pith have been colored brown.

Fig. 5 & 6. Cross-section through a protruding sporocarp. The intercellular mycelium (a) sends numerous haustoria (h) into the bark cells, penetrates abundantly between the outermost bark and epidermal cells and between the cuticle (c) and the latter (ee) develops *Spermogonia* sp. a large number of which covers the future urediospore fruiting body. This arises in the third layer of bark cells (st) in which numerous hyphae, thickened club-shaped towards the end (bb), penetrate from below between these cells and surround them on all sides. The spermatia produced by the spermogonia are enlarged at (s).

Fig. 7 & 8. A further developmental stage of a *Caeoma* fruiting body. The basidia arising between the partially reabsorbed and between the partially compressed bark cells (ff) in the beginning abstrict, by succession from the tip, thin-walled spores separated from one another by disjunctor cells (c). The disjunctor cells disappear with the formation of the spore wall (d). The bark layer covering the fruiting body still is not ruptured, since the number and formation of spores is still slight.

Fig. 9. Germinated spores.

Caeoma Laricis. R. Hrtg.

(Plate V, Fig. 10-16)

The Larch Needle Rust

A new fungus is associated with the numerous hostilities to which larch is exposed by weather, insects and fungi; to be sure, until now this fungus has been mentioned very rarely; however, it must be discussed here since a general distribution of it can occur or perhaps on other sites has already occurred. It is a new rust fungus closely related to *Caeoma pinitorquum*, which I first discovered on the needles of 3-40 year old larch in 1872, then discovered again this year in the forest botany garden at Neustadt.

Towards the end of May or in the first half of June one finds the fruiting bodies of this parasite on the needles (Fig. 10,11) and, to be sure, preferably distributed on their underside so that careful searching for them must be done if one will find them in a youthful condition. Later also the upperside of the needle becomes colored yellowish in the infected part and the needle shrinks together at least in the upper part; then often it is still barely possible to detect the dried up fruiting bodies on the needle which is colored similarly.

The mycelium is intercellular (Fig. 12-15), consists of hyphae 3-4.5 microns thick, which are richly branched, septate, and colorless. I have not been able to discover haustoria. The effect of the hyphae on the chlorophyll-containing parenchyma cells of the needle is very different in individual needles. Sometimes the green coloring material of the chloroplasts is gradually lost; these later lose their form and finally disappear completely. Only rarely did I see here and there a transformation of chlorophyll into starch. Often the green coloring material remains up to the time when, after the dispersal of the spores from the fruiting bodies, the entire cell tissue dies and one detects a detrimental influence on the health of the cell only in the withering of the tonoplast (Innenschlauch). Finally, the cell contents also take on a brownish-yellow coloration before the drying of the whole cell tissue occurs.

The mycelium spreads itself in the tissues of the needles over a distance of 5-15 millimeters, forms spermogonia preferentially on the underside, in part also on the upper side of the needle, which are detectable almost with the unaided eye as numerous small, longish swellings (Fig. 10, 11a). The structure of these corresponds almost completely with that of the spermogonia of *Peridermium Pini acicola* and *Caeoma pinitorquum*.

[88]

Mycelium penetrates here and there between the epidermal cells in abundant amounts; these are forced apart from one another and afford room for the development of the usually mostly flat cone-shaped organs (Fig. 12, 13, 15sp). The stroma, from whose upper side arise the slender hyphae which cut off spermatia, consists also here of a pseudoparenchyma (Fig. 13p), which is formed by the combined growth of numerous hyphae which penetrate outwards from the interior. Sometimes the spermogonia, which as a rule are covered by a thin cuticle, break through the epidermis in the shape of tufts projecting out for some distance (Fig. 13), which are detectable to the unaided eye as small, light points. The shape of the spermatia (Fig. 13b) is variable, pear-shaped or short cylindrical, rounded off equally on both sides.

The urediospore fruiting bodies appear almost without exception on the underside of the needle; its length extends 1-5 millimeters, its width is very variable, only rarely, however, reaching a third of the width of the needle so that when two fruiting bodies lie next to one another on each half of the needle, the mid-rib of the needle and the margin remain free (Fig. 11). Usually several small fruit bodies arise close to one another. If this fruiting body bursts, thus a white corroded margin projecting at right angles surrounds the shallow fruiting body nearly devoid of spores (Fig. 11b).

The stroma consists of numerous interwoven hyphae tangled up with one another, but extending mainly at right angles to the longitudinal axis of the needle. A cross-section through the needle and stroma (Fig. 14st) clearly shows the hyphae while in longitudinal section (Fig. 15st) the stroma appears as a pseudoparenchyma.

This formation of the stroma is easily explained by the orientation of the needle parenchyma. From the upper side of this stroma the small basidia arise (Fig. 14b) on whose tips occurs the abstriction of spores which are separated from one another by small disjunctor cells (Fig. 14c).

The short disjunctor cells disappear with the formation of the spore wall; they are differentiated from those of *Caeoma pinitorquum* spores by the smaller number, that is to say, the wider spacing of the wart-shaped thickenings. The spores are somewhat larger than those of *C. pinitorquum*, namely 20-22 microns large, roundish, egg-shaped or polygonal.

After lying in water for a longer time they enlarge considerably, up to a diameter of about 30 microns; the gold yellow plasma content is pressed against the outside by a number of bubbles filled with a hyaline liquid (Fig. 16).

I have not been successful in observing germination of spores. The ability of the basidia to cut off spores seems to be very limited. In any case in regard to the amount of spores produced by *Caeoma pinitorquum*, and above all by *Peridermium Pini*, it is very much inferior.

I have never seen more than six spores in a row, and the possible number must exceed this only slightly.

Each fruiting body is surrounded by a rather broad zone of sterile basidia, which grow forth into colossal many-celled tubes which force the epidermis towards the outside (Fig. 12, 14s). They form the projecting whitish boundary which is detectable to the unaided eye (Fig. 11) and which, prior to more exact investigations, I originally was inclined to interpret as the residue of a peridium. They take the place of the club-shaped so-called paraphyses which surround the little spore clusters of the genus *Epitea*.

Isolated yellow oil drops form the only contents of the otherwise colorless hyphae. Their length reaches as a rule 50-100 microns measured from the mutual stroma.

Of general scientific interest is the fact that often completely sterile fruit bodies occur, all basidia of which are deformed to such tubes (paraphyses) just as we have described as surrounding the fertile fruiting bodies. One part of such a nearly sterile fruiting body I have illustrated in longitudinal section in Fig. 15. The basidia, which lie on the periphery of the stroma, press against the epidermis in order to lift this off; they cannot develop completely uninterrupted into tubes but on the contrary form a large-celled pseudoparenchyma (Fig. 15e).

The basidia which lie more towards the interior and which hold the epidermis up from the spermogonium after it has been lifted off are not hindered in their development, grow forth to a 100-130 microns high colorless loose layer of pseudoparenchyma, whose end cells are mostly broadly rounded off (Fig. 15a) and which are formed just like the marginal tubes (illustrated in Fig. 14s).

However, often these are narrow and elongated (Fig. 15b) or almost spherical and have strongly thickened walls (Fig. 15c).

The latter then are scarcely different from the button-shaped thickened paraphyses in the hymenial layer of *Epitea salicis* which are illustrated by de Bary in Plate IV, Fig. 6, "Untersuchungen über die Brandpilze."

The entire sterile fruiting body is colorless and therefore with the unaided eye distinguishable from the gold-yellow colored, spore producing fruiting bodies.

Sometimes in the middle of the sterile cell tissue layer small spots occur with normal and fertile basidia, which then abstrict spores in the described manner (Fig. 15d).

After the dispersal of the spores the cell tissues of the needle die as far as the mycelium has developed in the interior; as a result of this the end of the needle, which has remained healthy up until now, is lost.

The parasite can cause noteworthy damage only when it damages in large amounts larch needles, about to the extent in which *Chermes Laricis* occurs, the damage of which is comparable with our fungus in numerous respects.

Until now the fungus still belonged to the mycological novelties. Indication of the alternation of generations is still lacking.

Noteworthy influences of the soil, weather, or general conditions of the attacked plant are not known. In a large nursery bed consisting of 3 and 4-year-old larch in their most vigorous growth there were a number of very vigorous about 10-year-old and about 40-year-old larch which bore the parasite.

Explanation of Illustrations

(Plate V, Fig. 10-16)

Fig. 10. End of a larch twig with three needles attacked by *Caeoma Laricis*.

Fig. 11. A piece of a needle (underside) attacked by *C. Laricis*. The spermogonia at (a) are distributed abundantly individually or in groups on the surface. The urediospore fruiting bodies (b) are bursting open; the epidermis is cast off; a narrow protruding margin remains only on the circumference of the fruiting body.

Fig. 12. Longitudinal section through a diseased spot on a larch needle. Three *Spermogonia* sp. are on the outer and underside of the needle, only one on the inner or upper side of this in cross-section. The intercellular mycelium develops a thick stroma under the epidermis, from which arise basidia which cut off rows of spores. On the margin of the fruiting body sterile basidia develop into tubes (s) which press against the epidermis.

Fig. 13. Cross-section through a spermogonium which has completely burst the epidermis and cut off spermatia which project in tufts enlarged more at (d). The hyphae originate from a stroma of pseudoparenchyma (p).

Fig. 14. Cross-section through a margin of a uredio fruiting body. The stroma (st) is formed of rather parallel running but entangled hyphae, from which arise small basidia (b) which cut off a few spores separated by short disjunctor cells (c). The mature spore (d) with yellow contents and with numerous wart-shaped thickenings on the surface. The basidia at the margin remain sterile and develop into many-celled large tubes (s) which press the epidermis at right angles towards the outside.

Fig. 15. Longitudinal section through a nearly sterile fruiting body of *Caeoma Laricis*. The stroma is sectioned at right angles to the longitudinal direction of its hyphae (st). Sterile basidia arise from it and grow up into large celled tubes, which, in the case of free development in the interior of the fruit body, terminate with large thick (a) or with elongated (b) or with button-shaped thick-walled end cells (c), while in the outermost part of the fruiting body, due to the pressure of the epidermis, they are transformed into a pseudoparenchyma (c) which consists of angular cells. Only in one small fleck (d) are the basidia normal and abstricting spores.

Fig. 16. (An error by the author. He means Fig. 14+ WM.) A greatly swollen spore after eight days in water and in moist air, the plasma of which is pressed against the margin by hyaline bubbles.

Peziza Willkommii R. Hrtg.

(Plate V, Fig. 16-21)

The Larch Bark Fungus

It is not my intention to relate my opinions and experiences about the disease of larch which was first worked on in detail by Willkomm, and which has been discussed so frequently in the past years; I will hold this for a later publication. On the contrary, it is a question only of me giving a single name to the fungus which is characteristic for the disease and which appears on the bark and especially on the canker spots, since, as I will show, until now this has been confused with other fungi and still possesses no name.

There occur several fungi which, with hasty examination, can be easily confused with one another and in which the fruiting bodies show a certain similarity with one another due to their cup-formed shape and the red coloration of the hymenial layer.

When Willkomm obtained the first material sent for the investigation of the larch disease, he sent diseased twigs with the fruit bodies of the larch fungus to Rabenhorst with a request for the identification of the fungus.

Since this famous mycologist certainly is favored with similar questions to an extent that it makes it impossible for him to base his determination wholly on the basis of microscopic investigation, thus this explains very well his error when he sent the larch fungus back to Willkomm with the name: *Corticium amorphum* Fr. It is not to be excused that the latter accepted this determination blindly, since one look at "Rabenhorst Deutschlands Cryptogamen-Flora" Vol. 1, page 391 must have informed him that *Corticium amorphum* Fr. or its synonyms *Thelephora amorpha* Fr. or *Peziza amorpha* Pers. belonged to the basidiomycetes and on the top of each tube bore four spores, while the larch fungus belongs to the ascomycetes. The fruiting body of *C. amorphum* (Fig. 16) is somewhat cup-shaped, to be sure, but with a shorter and broader stipe. The spores (Fig. 17c) arise on sterigma on the ends of large tubes (Fig. 17b), which are surrounded by paraphyses (17d) whose ends are pinched off like strings of beads.

The fungus is generally common, thus also in Neustadt in its forest botanical garden on the bark of killed white fir or on the twigs of this species.

Hoffmann in Giessen in the May, 1868, issue of Forst- und Jagd-Zeitung

with his report about Willkomm's "Mikroskopische Feinde" drew attention
to the fact that in the identification of the fungus an error must have occurred
since the fructification form indeed suited a *Peziza* but not a *Corticium*. If I
am not incorrect it was also Hoffmann who later placed the now accepted
name *Peziza calycina* Schum in place of *Corticium amorphum*. If this
taxonomy is now also incorrect, as shall be shown in the following, thus the
error was still quite excusable; indeed, in the identification of the fungus
Hoffmann could come to no other name, since the description of *Peziza
calycina* agrees fully with the character of the larch fungus. Fries in his
Systema Mycologicum, Vol. II, page 91, distinguished two varieties of *Peziza
calycina*, which are formed more according to the tree species than according
to the modification of the coloration of the fruit bodies, which in my opinion is
very unessential, namely *P. calycina α Pini silvestris*: in ramis dejectis
putrescentibus; *Pini silvestris β Abietis*: ad cortices abiegnos.

In addition, there occurs in Fries' Elenchus fungorum, Vol. II, page 8, a
third variety, *Peziza calycina γ Laricis*: in ramis *Pini Laricis, P. balsameae*.
Rabenhorst (Deutschlands Cryptogamen Flora, page 362) names only the
first two varieties; he differentiates these, however, only according to the
coloration of the fruiting bodies.

I have illustrated the *Peziza calycina* Schum *β Abietis*, which I collected in
the Ore Mountains on the bark of dead white fir twigs (Fig. 18, 19). Fruit body
and the shape of the asci and paraphyses are different only in unessential
points from the fruit bodies (Fig. 20) and the asci and the paraphyses Fig. 21)
of the larch disease fungus.

A type of basic difference, however, lies in the completely different sizes of
the asci, spores and paraphyses. Therefore I have named the fungus of the
larch disease so exactly described and so correctly illustrated by Willkomm in
honor of him: *Peziza Willkommii*. The difference is the following *P. calycina*:
size of the asci: 66 microns, the spores: 7.5 microns, the paraphyses: 75
microns; *P. Willkommii* size of the asci: 160 microns, the spores: 18 microns,
the paraphyses: 200 microns.

In addition there also occurs a difference in the form of the spores which in
P. calycina is often egg-shaped (Fig. 19a), in *P. Willkommii* always completely
elliptical (Fig. 21a). The shape of the fruit bodies is too variable to serve as a
suitable criterion for the differentiation of the two species. About the
remaining characters of *P. Willkommii*, I refer to the illustration given by
Willkomm from page 200 forward in the "Mikroskopische Feinde." The sizes
of enlargement listed for Plate 14 of the illustrations of that report are too
large.

It seems probable to me that *P. Willkommii* is identical to *P. calycina* var.
Laricis Chaillet.

It is important to note, that the fungus is characteristic of larch because the
general occurrence of *Corticium amorphum* and *Peziza calycina* throughout
Germany for a long time was a reason which could be used by the opponents
of Willkomm's opinions to absolve the larch fungus from any complicity in
the spread of the larch disease for a few decades.

[93]

Explanation of Illustrations

(Plate V, Fig. 16-21)

Fig. 16. Section through a fruiting body of *Corticium amorphum* Fr. on white fir bark. (a) The hymenial layer.

Fig. 17. A number of basidia (b) from the hymenial layer of *Corticium amorphum*. On the ends of these arise four outgrowths (sterigma), whose end swells into a spore (c). Paraphyses (d) which are thickened bead-like towards the top and sometimes are branched.

Fig. 18. Section of a fruiting body of *Peziza calycina* on white fir bark. (a) The hymenial layer.

Fig. 19. A number of asci (b) from the hymenial layer of *P. calycina*, in each of which eight egg-shaped spores (a) have formed, with numerous thread-like paraphyses (c).

Fig. 20. Section through a fruiting body of *Peziza Willkommii* on larch bark. (a) Hymenial layer.

Fig. 21. Several asci (b) from the hymenial layer of *P. Willkommii* in each of which eight elliptical spores (a) have formed with numerous thread-shaped paraphyses (c).

Hysterium (Hypoderma) macrosporum R. Hrtg.

(Plate VI, Fig. 1-17)

The Spruce Needle Blight

Causal Organism of Spruce Needle Browning, Needle Reddening, and Needlecast

In the winter of 1868/9 I received spruce twigs sent in by Count Bethusy-Huc from the forest district of Morbach in the management district of Trier which were attacked by a disease which I have found since then distributed throughout all of north and central Germany and, to be sure, to an extent which deserves the greatest consideration of the foresters, since it must be at least as frequent and injurious as the so-called spruce needle rust which is caused by *Chrysomyxa Abietis.*

Needle browning is distinguished at first glance from needle rust, with which until now it may have been commonly confused by foresters, in that in the case of the latter the needles of one-year-old twigs die, never those on shoots of any older ages, while in the case of needle browning the needles brown and redden without taking on at any time a light yellow or goldish yellow coloration as is characteristic of the needles attacked by *Chrysomyxa.* The parasite which causes needle browning or needlecast of spruce has not been described by mycologists until now since, without closer examinations it has been considered to be identical with the *Hysterium (Hypoderma) nervisequium* De C. on white fir.

The exact description of the latter named fungus will follow in the next chapter.

I compile here the evidence from the literature which demonstrates that the spruce needle blight in question has still not been described, that is to say, has been confused with *Hypoderma nervisequium.*

De Candolle* first described *Hypoderma nervisequium* and said of it: "Cette espéce croit á la surface inferieure des feuilles du sapin, elle se developpe sur la nervure moyenne, d'abord par des points oblongs et interrompus, ceux ci se réunissent, tous ensemble et forment une raie

---

*De Candolle. Flore francaise V. 167.

[95]

longitudinale, convexe, noirâtre, qui occupe toute la longueur de la nervure, et qui s'ouvre par une fente longitudinale, la substance interne est de couleur pâle, non pulvérulente, les individus âgés offrent souvent une petite cavité longitudinale. Cette production a été observée dans les Vosges par MM. Mougeot et Nestler".

Fries[**] saw *Hypoderma* on spruce (*Abies excelsa*) and considered this to be a variety of *Hypoderma nervisequium,* of which he says in his Systema mycologium: "In foliis *Pinus piceae*".

Rabenhorst[***] gives a description of *Hypoderma nervisequium* De C. which in its inexactness can pass in all cases for both varieties: "Perithecia on the lower needle surface along the vein bound together in very long, straight, black, slightly arched stripes, which open themselves with a common longitudinal crack and the pale fruit body lying bare....on the needles of the *Pinus picea* (Linne or Du Roi?) in the Vosges Mountains, in upper Italy".

Duby[****] describes *Hypoderma nervisequium* with the following words: "hypophyllum nervisequium innatum lineare demum in striam longissimam rectam confluens nigrum nigro fuscescensve, labiis convexis arctissime conniventibus demum lineam rectam relinquentibus et acie interdum expallentibus thecis late clavatis sessilibus paraphysibus filiformibus brevioribus sporas hyalinas filiformes apice paulisper incrassatas homogenas flexuosas inordinate dispositas foventibus. Ad folia Abietis et Piceae, in Vogesis et Germaniae. Primo interruptum demum confluens in striam longitudinalem totius nervi longitudinem occupantem."

From the previous description of *H. nervisequium* it appears that Duby held the parasitic fungus on spruce to be identical to that on fir.

Fuckel[*****] named the parasite in the text and in the index of his second appendix to Symbolae mycologicae: *Hypoderma nervisequium* Fckl. (instead of De C.); described as Fungus spermogonium of this parasite a fungus which he named *Septoria Pini* Fckl., with the following diagnosis: "Peritheciis immersis, in acervulis lineari oblongis, seriatis, per epidermidem demum erumpentibus; cirrhis candidis; spermatiis oblongoclavatis, uniseptatis, hyalinis. On living needles of *Pinus excelsior,* frequently in the fall. Causes the shedding of green needles. The *Fungus ascophorus* belonging to this species occurs rarely in the spring on the lower surface on the dead but still attached needles of *Pinus excelsior.*"

Since in his description of our forest trees Fuckel made use of the most peculiar names, and it remains doubtful whether *Pinus excelsior* is understood as *Abies excelsa, Abies pectinata,* Himalayan pine (*Pinus excelsa*) or any other plant, thus an interpretation of the previous publication is very difficult. I have not observed a fungus on spruce, fir or pine which possesses any similarity to *Septoria pini* Fckl.; in any case I must doubt that this fungus bears any relationship to *Hypoderma.*

[**]Fries: Systema mycologicum, Vol. 2, page 587, 1823. Elenchus fungorum, Vol. 2, page 144, 1828.
[***]Rabenhorst: Deutschlands Cryptogamen Flora, page 156, 1844.
[****]Duby: Memoire sur la tribu des Hysterinees. 1860.
[*****]Fuckel: Symbolae mycologicae, 1869, and appendix II, 1873.

[96]

In the second supplement, page 51, Fuckel says about *Hypoderma nervisequium* Fckl.: "The mature tube-bearing fungus I found in Mittelheimer Forest, Frankensteiner Kopf, on the leaves of *Pinus Picea* (Linne or Du Roi?) in the spring but only on one tree, but on this very abundantly.

The fungus attacks the living needles, whereby these rapidly become yellow colored which gives the tree a colorful appearance, in that these yellow needles remain hanging on for a while; later they fall off and the fungus becomes completely mature. Without doubt it is very injurious to the tree, since the tree is defoliated".

The statement of De Condolle is decisive: *Hypoderma nervisequium* comprises the species which occurs on fir (*Abies pectinata*). Since I will prove in the following the specific difference of the species which is parasitic on *Abies excelsa,* thus there follows from this the justification to give this fungus a new name. It is based on a mistake that Thumen has published in his *Herbarium mycologicum* under the name of *Hypoderma longisporium* R. Hrtg. the samples submitted by me.

THE MYCELIUM

The mycelium of *Hypoderma macrosporum* develops intercellularly in the leaf parenchyma of spruce needles and, it appears, without sending haustoria into the cells, which, however, very soon shrivel up if they come into contact with hyphae (Fig. 5 n.o.). The latter are of varied thicknesses, from 1-16 microns in diameter. In the first stage of the disease the 10-16 micron thick hyphae predominate; from these the thinner and quite fine hyphae arise at irregular intervals here and there in large numbers (Fig. 6). The further the destruction of the cell tissue has progressed, the more the thick hyphae disappear until at last they are completely replaced by fine hyphae.

In the most youthful condition (Fig. 6a) the hyphae except the outermost thin membrane are filled by a coarsely granulated plasma which is colorless and which is colored yellow by iodine. Very soon one detects, however, a double layering of the cell wall and the formation of septa (Fig. 6b).

The primary cell wall, which is colored yellowish by iodine, reaches at most a thickness of 0.5 microns; in the interior of this is formed a second cell wall (Fig. 6d), whose thickness often exceeds 5 microns. This is distinguished by clear layering and by the intensive clear blue coloration on treatment with iodine. Towards the lumen of the hypha it is bordered by a third layer identical to the primary wall, which is distinguished especially clearly by the yellow coloration when the secondary cell wall is colored blue by iodine (Fig. 6e).

The lumen, which often forms only a very fine canal, in the beginning still contains plasma which at last disappears except individual granules or oil drops (Fig. 7b).

The chemical composition of the cell wall of the hypha undergoes a change after a certain time, which expresses itself at first in that the middle wall is no longer colored blue by iodine, so that browning of this and finally a complete decomposition and destruction occurs.

The thinner hyphae are formed just like the thick ones, only the individual layers do not appear so clearly; in quite fine hyphae they are not detectable at all. At a certain stage, however, they are likewise colored bluish by iodine.

THE FRUIT BODIES AND REPRODUCTIVE ORGANS

Sometimes already after two months, in other cases only half a year after the disease and browning of spruce needles, on the red needles attached very firmly to the twigs the formation of the perithecia which produce asci begins and to be sure as a rule only on the two undersides of the needles.

These appear as numerous small, somewhat long, dark flecks (Fig. 2b) of which those standing closest together generally coalesce into long, black stripes (Fig. 3b) which almost never extend the entire length of the needle.

These originally appear only slightly above the surface of the needle; in late fall these first become arched and open through a common, sharp longitudinal split mostly in the months of April and May during moist weather. From the whitish hymenial layer, which thereby is freely exposed, the spores are expelled (Fig. 10). The empty perithecia are distinguishable with the unaided eye from those which have not yet ruptured by the fine split on the ridge of the longitudinal black swelling which is still detectable after the shooting of the perithecia (Fig. 4).

The formation of the perithecia is the same in both *Hypoderma macrosporum* and *H. nervisequium.*

The first stage of development (Fig. 21a for *H. nervisequium*) is illustrated in Fig. 7. From the interior of the needle, where the perithecium will be formed, there appear fine mycelial threads (Fig. 7a) in the epidermal cells (Fig. 7f), which develop there into a fungus mass which is granulated at the first glance (Fig. 7c) and by which the epidermal-cell layer is split apart and the upper half, together with the cuticle, is forced toward the outside, the lower half forced somewhat toward the interior. Close examination of the granulated mass which fills the lens-shaped cavity which thus arises shows that it consists of hyphae whose individual cell members are more or less roundish, so that the tightly-compacted hyphae appear as a mass composed of roundish cells (Fig. 7c). In the beginning this is completely colorless, so that the earliest stage is not detectable by a brown coloration of the needle; soon, however, the part of the fungus mass lying towards the outside is colored dark brown (Fig. 7d) and becomes the firm perithecial wall (Fig. 10b) which protects the hymenial layer which arises underneath it. In the middle where the perithecium is arched the most, it attains a thickness of 100 microns, on both sides it tapers gradually until it entirely disappears on the border of the perithecium. With the formation of the perithecial wall the formation of the hymenial layer begins. Where the granulated fungus mass first becomes loose from the lower part (Fig. 7g), one observes that the uppermost cells of the latter lengthen (Fig. 7e) and grow up into parallel hyphae which force the perithecial wall towards the outside (Fig. 8p).

These are about 4 microns thick, rich in plasma, and surrounded by only a very fine jelly layer; later, they are converted into paraphyses.

In Fig. 9 they have already changed insofar as the jelly layer has thickened itself significantly at the expense of the plasma. The lumen which contains the plasma is reduced to a very fine canal, and with hurried examination this transformation can be compared with the transformation of the mycelial threads (Fig. 6).

The paraphyses, however, lack the sharply delimited double contoured outer layer; the jelly layer is not stratified, is not colored blue by iodine, is severely attacked by potassium hydroxide but not completely dissolved; on the other hand with suitable lighting one frequently detects on the circumference of the hypha an extremely thin jelly layer. This bears more the character of a jelly-like secretion of the cell wall than a jelly-like cell wall, since a sharp outer boundary is detectable neither in the paraphyses nor in the jelly layer of the spores (Fig. 12). The lengthening paraphyses are hindered in their free development by the lack of space and bend wavily to and fro. On the circumference of the perithecium where the space between the stroma and the perithecial wall is the slightest, the paraphyses cannot develop at right angles to the perithecial wall but shove themselves between the perithecial wall and the paraphyses which arise more towards the interior, often winding and, bending inwards, likewise to some extent forcing their way towards the opening which arises later (Fig. 9 at the right).

From the stroma the young tubes (asci) in which the spores develop (Fig. 9as) arise between the paraphyses. One detects originally only a single nucleus in them, at further stages of development two, four, and finally eight nuclei as a result of numerous repeated divisions (Fig. 11b.c.d.).

The asci do not arise simultaneously altogether; one frequently sees larger or smaller asci, later completely ripe asci next to quite young asci (Fig. 11).

Before the maturation of the perithecium a further change of the paraphyses occurs.

While they still lengthen, they abstrict on the tips long, rod-shaped organs (Fig. 11k), which can be considered to be spermatia. It is doubtful to me, however, whether all paraphyses abstrict such rods or only particular ones; in general they are not frequent. One could consider these as fragments of the paraphyses if their origin on the tip of the latter was not recognizable with great certainty. In the case of *H. nervisequium* these occur also. The tip of the paraphysis thickens, often finally completely spherical or only club-shaped. A side branch develops not uncommonly from the thickened ends (Fig. 11a.a.) or a small oval cell is abstricted from the tip (Fig. 11m) whose development I had the opportunity to observe. In many paraphyses the jelly layer disappears; in others it remains and often binds together the majority of these so that they can be separated from one another only with force.

In the tubes (asci) arise cylindrical spores (Fig. 12) whose length reaches about 60 microns which affords me the choice of the species name. They are uniformly thick towards the top, scarcely somewhat thickened at the apex; towards the bottom they gradually come to a point. The content is colorless, coarsely granulated plasma; a jelly layer surrounds the spore externally in many cases; however, also very often it is absent. The development of the spores from the nuclei is shielded from observation in the plasma-rich interior of the asci.

One almost always detects them, when they have already reached a significant size, by the nuclei in their contents which are arranged in rows (Fig. 11e.e.). The mature spores extend usually from the apex of the ascus to the base, the eight lie usually parallel to one another (Fig. 11g), or they also are twisted spirally in the interior of the ascus (Fig. 11f). They escape the asci through a small opening in the tip which is formed only by the spores (Fig. 11h), which is very clearly recognizable in empty asci (Fig. 11i), or the asci are split open by the swelling by the spore jelly, whereby almost always an upper part with a sharp, straight margin is loosened from the larger or smaller lower part (Fig. 23, left). If one moistens under a cover glass dry prepared sections through a mature perithecium, thus the asci empty themselves through the tip or by splitting with great energy and apparently the swelling of the jelly layer is the cause of the emptying. In many cases the asci loosen themselves entirely from the stroma and the spores come out of the lower opening which thereby arises (Fig. 12d); thus must I leave undecided whether this tearing open of the asci results only from injury in the preparation of the material, or whether the swelling of the spores in the interior can bring about the separation from the stroma.

The germination of the spores in water under the coverglass on a microscope slide is complete after 24 hours (Fig. 12a) in that in most cases near the thick end, however at times also near the thin end, a germ tube grows forth, which does not quite attain the diameter of the spore itself. After 48 hours the tube had reached the length illustrated in Fig. 12 b.c. Sometimes the spores germinate already in the interior of the asci; the germ tubes bore through the wall of the ascus and branch repeatedly (Fig. 12c). After 96 hours the tubes had attained the length illustrated in Fig. 12c, whereby they perished due to the influence of bacteria which had colonized in the meantime. The described process of germination was noticed with a culture undertaken on May 14, 1872.

If the perithecia are mature, they remain closed for a long time until long persistent rainy weather, by which the killed needles are completely wetted through, causing the swelling of the organs in the interior of these and by which the bursting of the perithecial wall is brought about (Fig. 10).

The longitudinal swelling, which arises from numerous perithecia arranged in rows with one another and also by perithecia which coalesce in the interior, splits open and shows the whitish hymenium until the spores are disseminated or until further dry weather occurs and the organ again closes before liberation of the spores. In dry years it can easily happen that many organs do not open at all and then the spores die in the interior of the asci.

I have often encountered still unopened perithecia long after the normal period of maturation of the spores, in which the spores were found in a condition which justified the supposition that they had already died.

The needles with the empty perithecia remain still attached to the twigs for several years until they are nearly decomposed.

Usually the formation of small organs (Fig. 2a) which we will designate temporarily as spermogonia of the fungus precedes the development of the perithecia. In the case of *H. nervisequium* quite analogous organs regularly occur in large numbers on the upper side of silver fir needles (Fig. 20, 21b) (Ore Mountains) or they arise individually and scattered in the case of spruce needles especially on the lower side. They become dark colored not at all or only after death.

These spermogonia (?) like the perithecia arise by the splitting of the epidermal cells away from one another and at first are scarcely different from the condition illustrated in Fig. 21a.

However, no wall occurs; on the contrary, parallel extremely slender hyphae arise from the stroma (Fig. 22), which press the epidermis towards the outside and abstrict numerous longish, elliptical spermatia(?) (conidia?) from their tips.

Sometimes the perithecia develop normally up to the stage illustrated in Fig. 7; paraphyses do not develop from the stroma (Fig. 7c), but instead a thin layer of hyphae which abstricts the same spermatia or conidia before the organ has burst. Or there develops under the epidermis from the granulated mass of the wall a cushion consisting of slender hyphae which breaks through the epidermis and appears externally. Finally fungus cushions (Fig. 14 n.m.) also form in the wall of the perithecia after these are already empty, which appear on the outside in moist weather in a reddish-white cushion (Fig. 4a.a.).

The surface of this cushion (Fig. 15) consists of bristle-shaped pointed hyphae, which abstrict numerous longish elliptical cells. These germinate very easily in water, after they have enlarged not inconsiderably (Fig. 16) and develop into hyphae (Fig. 17 eight days after sowing), which on the tips of hyphal branches by successive abstriction form large amounts of conidia of the same shape, but of greater size.

The newly arising conidia shove the previously formed ones to the sides, since they all remain clinging to one another, thus small, easily-broken-down spore clusters are formed. The organs which form conidia which are able to germinate and which arise in so many different ways I have not allowed to remain unmentioned, since they are such regular associates, predecessors, or followers of the ascus-bearing perithecia; still I will not deny the possibility that they are an extraneous saprophyte which has colonized the needle killed by *Hypoderma* and which can form its conidial cushions in so many different ways.

With the form of the disease which is to be described later as needlecast, organs appear next to the perithecia which are distinguishable from the previously described ones only by the odd shape of the abstricted cells (Fig. 13 a.b.).

MODE OF LIFE OF THE PARASITE

Healthy, at least one-year-old spruce needles always show chlorophyll and the cells are very rich in starch in the leaf parenchyma only in the spring up to

June or July. As we will see soon, the progress of the disease is often different in several ways according to the climatic differences of the locality, the infection of the needles occurs in the mountains in May, at Neustadt-Eberswalde only in September or October.

This difference produces an interesting explanation about the effect of the fungus mycelium on the tissue of the needle, which is killed by it. If the disease occurs in the fall at a time in which the cells contain only chlorophyll, thus these cells wither soon after contact with the hyphae which grow in the intercellular spaces, the green coloring material disappears in a short time and the brownish collapsed mass contains only a few granules, destroyed chloroplasts, which are not colored by iodine (Fig. 5o.).

If the needles become diseased at a time when they contain starch, such as in May, thus likewise the cells which are killed by contact of the hyphae wither up but, however, contain starch granules for a long time in an apparently unchanged condition.

From May until October of the same year the cells full of starch are colored intensely blue by iodine (Fig. 5m).

From October onwards individual cells begin to empty until in May of the next year every trace of starch disappears from the needles; it is used up by the fungus mycelium.

If infection occurs at a time when the transformation of chlorophyll into starch has begun in the healthy needles and scarcely can be detected by iodine, then contact of the cells with the fungus mycelium results in a sudden transformation of the chlorophyll to starch.

In these needles one finds that as far as the hyphae have spread out from the infection spot, all cells are crammed full of starch while in the still healthy needle part often scarcely a trace can be detected. Here must we conclude that almost all conditions were present for the transformation of the chlorophyll into starch, and it only needed the stimulation by the hyphae of the fungus to complete the transformation.

Whether this stimulation is based on a withdrawal of nitrogen etc. by the nutritional process of the fungus or in any other process must remain undecided.

The symptoms of the disease, which are produced by the parasite, are of three different types.

1. In the Ore Mountains, in the Harz, in the Weser Mts., Braunschweig, and so forth, before the beginning of new shoot formation in May there occurs a discoloration on more or less numerous needles, rarely on all, often on only individual needles, of the previous year's shoots, which begins on the tip or the base or the middle of the needle, and gradually spreads over the entire needle.

The fresh green color is transformed into a dirty dark green. Also individual needles of the two- and three-year-old shoots are discolored, and like the one-year-old shoots become gradually reddish-brown.

The newly diseased needles show on the border of the healthy and the brown part the form which is illustrated in Fig. 5 m.n. In July the spermogonia and perithecia arise (Fig. 2,7) which by August attain the stage illustrated in Fig. 8. In October the asci arise (Fig. 9); they persist, however, in an immature condition during the winter, until in April of the next year maturation occurs (Fig. 10,11). In April and May the asci empty when, after several days of rainy weather, the needles have been thoroughly wetted and the perithecia have burst forth. In June one encounters mostly only empty perithecia on the needles which remain firmly attached for several years.

Since the disease does not limit itself to the needles of the previous year's shoots, individual needles of older shoots also are attacked likewise, thus in May before bud break on the two- and three-year-old shoots between the healthy needles one can encounter some needles which are freshly diseased, some which bear ripe perithecia, and some which bear empty perithecia from earlier years.

Usually the diseased needles remain attached to shoots and develop the perithecia, that is to say, the spermogonia in the illustrated manner. Frequently, however, a larger part of the brown needles falls off in summer before the formation of the reproductive organs, so that a limited type of needlecast occurs.

The diseased needles appear in the first year completely filled with starch which is completely consumed by the next spring, before the maturation of the spores.

2. At Neustadt-Eberswalde the progress of the disease is quite different. The browning of the needles occurs in the two-year-old shoots in September, October or sometimes later during the winter without being accompanied by the formation of starch.

In June of the next year only the perithecia arise on the needles, which in August reach the stage illustrated in Fig. 8, in October the stage illustrated in Fig. 9. Maturation and dissemination of the spores occurs in April and May of the third year of the disease.

3. In the case of the two previously described disease forms only a limited portion of the browned needles fall off before the formation of the perithecia; a third form of the disease which I observe on numerous spruces in the Neustadt Arboretum is distinguished by the falling of all diseased needles a short time after the disease. The browning of the needles begins on all shoots in August; in the course of the winter or already in late fall almost all these fall, so that only a few remain attached to the plant. Therefore the name needlecast (Nadelschütte) must be designated. There is a total or partial casting according to whether all needles of all shoots fall off or only a part of these die. In the latter case the remaining green needles show numerous small brown flecks.

In the first case there remain on the shoots mostly only such needles as are diseased only on the tips, whose bases thereby remain green; a profuse resin impregnation on the boundary between the healthy and diseased portions of the needle has hindered the further development of the mycelium.

On the diseased tips of these needles there develop already in the fall the spermogonia, illustrated in Fig. 13, and the rudiments of the perithecia which

[103]

here, however, as a rule, stand more isolated and do not coalesce into longitudinal swellings. In March there occurs the stage illustrated in Fig. 8; in June and July the spores are mature and are disseminated.

If the casting of the spruce repeats for several years on the same tree as I had opportunity to observe in Neustadt, then the life of the tree can become easily endangered by this disease.

The progress of the three forms of the disease becomes clearer by the following summary. This is based on numerous investigations which were carried out for several years, and in it the progress of development of a needle from formation up until the emptying of the asci is indicated as follows: gr means green healthy; br means reddish-brown coloration without perithecia; 7, 8, 9, 10 the stage of development of the perithecia which is designated on plate 6 by the same numbers; x signifies the needle with empty perithecia; - - - signifies the defoliated condition.

Month	I. Needle Reddening Ore Mts., Harz			II. Needle Reddening Neustadt-Eberswalde				III. Needle Cast Neustadt-Eberswalde	
Age of Needles	0/1	0/2	2/3	0/1	1/2	2/3	3/4	0/1	1/2
January	.	gr.	9	.	gr.	br.	9	.	−u.7
February	.	gr.	9	.	gr.	br.	9	.	−u.7
March	.	gr.	9	.	gr.	br.	9	.	−u.8
April	.	gr.	10	.	gr.	br.	10	.	−u.8
May	.	br.	10	.	gr.	br.	10	.	−u.9
June	gr.	br.	x	gr.	gr.	7	x	gr.	−u.10
July	gr.	7	x	gr.	gr.	7	x	gr.	−u.10
August	gr.	8	x	gr.	gr.	8	x	br.	−u.x
September	gr.	8	x	gr.	gr.	8	x	br.	−u.x
October	gr.	9	x	gr.	br.	9	x	−u.br.	−u.x
November	gr.	9	x	gr.	br.	9	x	−u.7	u.x
December	gr.	9	x	gr.	br.	9	x	−u.7	−u.x

The previous summary gives the different stages of development in which the disease occurs for each month of the year. Fig. 1 demonstrates a spruce twig, from which the one-year-old and the tips of the two-year-old shoots have been removed so that one sees a part of the three-year-old and the two-year-old shoots, taken in December from the Neustadt Arboretum. The one-year-old needles are green (compare column 2 of the summary), the two-year-old needles are in part reddish-brown (Fig. 1a), the three-year-old needles (Fig. 1b) bearing still not fully mature perithecia (Fig. 9).

That newly diseased needles also can form on three-year-old and older shoots should be emphasized again.

The differences in the progress of the disease between column one and column two are explained by the favoring of fungus development by the moist climate in the mountains as opposed to the dry air at Neustadt.

In the case of column 1 the disease of the previous year's needles in May is followed immediately by the dissemination of the spores in April and May. The moist air promotes not only the germination of the spores and thereby the infection, but also the development of the reproductive organs on the diseased needles.

These arise in the case of column 1 already in July of the same year; because of the beginning of winter maturation is delayed until April of the next year.

At Neustadt, column 2 of the table, the perithecia likewise mature in April and May. Dissemination of the spores, however, appears to be hindered by the dry air in the summer months and likewise germination appears to be hindered and the browning of the needles begins only in October, with the appearance of moist air. Since the winter immediately follows the disease, the formation of perithecia occurs only in June of the next year but yet too late to reach maturity in the same year, which occurs exactly one year later than in the case of column 1.

Regarding the spread of the disease, I have observed it to be very intensive in the Ore Mountains at Neustadt-Eberswalde where needle reddening has attained a very noticeable degree.

In the Altenberg forestry district in the Ore Mountains of Saxony it occurs at an elevation of 700 to 900 meters on 10- to 40-year-old spruce and quite generally in individual beech stands with spruce undergrowth; in pure well-growing spruce stands, however, it is limited more to individual very well growing spruce, which have taken on an intensive red brown coloration.

In the Hermsdorfer forestry district in an approximately 30-year-old stand mixed with beech it occurred so intensively that about 10% of all the spruce have been reddened.

In Neustadt in the Institute Forests as well as in the Arboreta the disease has become worse each year since 1868, when it first appeared to a noticeable extent, so that an increase which progressed to the same degree would be extremely serious.

On the southern border of the Harz Mountains the disease is somewhat more widespread than I noticed in the northern boundary at Harzburg.

In Solling until now only individual trees in 10- to 25-year-old stands have been reddened. These are mostly quite strongly developed trees.

In the Trier district, Morbach forest district, the disease should be not uncommon especially on moist sites, etc.

On older diseased trees the uppermost part of the crown as a rule is spared; all facts indicate, therefore, that moist air favors the development and spread of the fungus, also that the lower and inner portions of the tree crown less exposed to air movement are more endangered.

The parasite occurs very often with *Chrysomyxa Abietis* on the same tree or at least in the same stand and the external conditions upon which the development of two parasites depends seem to be rather the same.

Explanation of Illustrations

(Plate VI, Fig. 1-17)

Fig. 1. The underside of a diseased spruce twig in winter (Neustadt). The base of the two-year-

[105]

old shoot shows several killed brown needles (a); the three-year-old shoot possesses needles which are partially green and healthy and partly newly diseased and brown (c), mostly, however, needles with perithecia which coalesce into black stripes (b).

Fig. 2. Browned spruce needle with ripe spermogonia (aa) and newly developed perithecia (bb).

Fig. 3. Browned spruce needle with mature perithecia which still have not ruptured through (b).

Fig. 4. Browned spruce needle with empty perithecia. Spermogonial cushions (aa) breaking through the epidermis or penetrating out of empty perithecia.

Fig. 5. Longitudinal section through a recently diseased spruce needle which shows at (m) still completely healthy, chlorophyll-containing parenchyma. The chloroplasts are just beginning to transform themselves to starch granules. In the intercellular cavities occur thick and thin plasma-rich hyphae. Through the effect of these hyphae the chlorophyll is transformed suddenly into abundant starch (n). If the disease occurs in the fall, then no transformation to starch occurs, the cells shrivel together, the contents form a brown formless mass (o).

Fig. 6. A very youthful hypha with thin cell walls and granular colorless plasma (a). Thick and thin lateral hyphae arise behind the growing point, at (b) cross walls arise, (c) is the mature primary wall. At the right a part of the older mycelium. Within the primary wall (c) there has formed a layered, thick secondary wall which is colored blue by iodine, which towards the narrow, plasma-containing lumen is bounded by a third wall layer which is colored yellow by iodine. At (f) distinct cross walls.

Fig. 7. Part of a young perithecium in cross-section. The thinner hyphae (aa) already outnumber the thick hyphae (b) and grow in the epidermal cells (f) which are split apart by the abundant development of the mycelium so that between the upper and lower part of the epidermis a narrow lens-shaped cavity is formed. This is filled by a granular fungus mass (c) which consists of short-celled hyphae. The outer part (d) of the brown mass dies next, becomes colored dark brown, and forms a layer which protects the hymenial layer. From the lower part of the filling layer the young paraphyses arise (e).

Fig. 8. Further stage of development of the perithecium. The covering (d) is unchanged, but forced towards the outside by the paraphyses (p) which grow up parallel to one another and at right angles to the stroma (e).

Fig. 9. Further stage of development. The paraphyses (p) are surrounded by a jelly layer, while the plasma-containing lumen of these has become very narrow. From the stroma (e) the young asci arise (as) between the paraphyses.

Fig. 10. A mature and ruptured perithecium in cross-section. The perithecial wall (d) is split down the middle. At (e) the stroma from which arise numerous asci, some already empty, some bearing mature spores, some immature. Individual ejected spores.

Fig. 11. A part of the hymenial layer enlarged. The paraphyses (p) in part abstrict rod-shaped organs (spermatia?) (k), partially roundish cells (m) from the ends, but mostly at maturity are thickened at the ends either spherically or club-shaped, sometimes are jointed or bent (aa). The asci are in part still undeveloped, containing one (b), two (c) or eight (d) roundish nuclei. Immature (ee) and mature spores, which are arranged mostly parallel (g), at times spirally (f) and nearly fill up the lumen of the ascus. The plasma disappears completely or to only a small trace. The ejection of the spores occurs from an opening on the tip (h) which forms at maturity, or the asci split further down and the spores appear from the base (Fig. 12d). The empty asci shrink together (i).

[106]

Fig. 12. Germinated spores with and without a jelly layer. The germ tube appears partly near the thick end (a), partly somewhat below (b.b.). Sometimes germination occurs within the ascus (c.c.). Germ tube 96 hours after sowing of the spores (e).

Fig. 13. Spermogonium with cylindrical spermatia (b) from a spruce needle.

Fig. 14. Empty perithecium, in whose perithecial wall arise reddish-white spermogonia or conidial cushions. (s) conidia or spermatia.

Fig. 15. Surface of the conidial cushion enlarged.

Fig. 16. Germinating conidia.

Fig. 17. Sporulating fungus hypha which has originated from the germinated conidium.

Hysterium (Hypoderma) nervisequium D.C.

(Plate VI, Fig. 18-25)

The Silver Fir Needle Blight

Causal Organism of Silver Fir Needle Browning, Needle Reddening, and Needlecast

For about five years a disease of silver fir has steadily spread in the forest arboretum in Neustadt which can be designated as needle browning and needlecast, and which, in the Ore Mountains, especially in the Olbernhau district, both in pure fully stocked fir stands as well as in fir stands mixed with beech, has attacked so severely that they have attained a very sickly appearance. The parasite, to which this disease is attributed, *Hypoderma nervisequium*, is so similar to the causal organism of spruce needle browning, that it may suffice to emphasize only those characteristics by which this parasite is separated from the former.

The mycelium is formed completely similar to *H. macrosporum*, only in the first stage of the disease the fine hyphae predominate over the thick hyphae so that the latter can be characterized as rare.

THE FRUIT BODIES AND REPRODUCTIVE ORGANS

The perithecia which produce the asci form on the underside of browned needles firmly attached to the twigs in great numbers on the veins which pass through the middle of these needles, and the perithecia, as a rule, coalesce into a common, somewhat undulating black stripe which reaches almost the entire length of the needle. However, this often is interrupted by one or more gaps (Fig. 18, 19).

In those needles which fall soon after infection by the parasite, single perithecia form on the upper side as well as the lower side of the needle on the soil. Often these are likewise arranged in rows; usually mostly they are completely scattered.

The upper side of the retained needles in Neustadt remains almost without exception without any fruiting bodies, while in the moist air of the mountains, numerous coalescing spermogonia form a broad, dark band which covers the middle of the needle (Fig. 20).

The spermogonia always precede by some time the formation of perithecia on the underside.

At Neustadt, the spermogonia arise only on those needles which are cast soon after browning in the summer. Lying on the soil, these develop abundant spermogonia, especially on the underside of the needle. However, these do not coalesce with one another, but remain isolated exactly as the spermogonia on the needles of spruce.

In regard to the development of the perithecia (Fig. 21a), I referred to this in the previous section.

Differences first occur at or shortly before maturation. These consist first in the formation of paraphyses which never thicken at the apex, never are branched or bent, but likewise cut off rod-shaped swellings (spermatia?) (Fig. 23k). Rare, but most peculiar are the pear-like swellings which arise on young paraphyses below the tip (Fig. 25a) and due to the lengthening of the latter are formed into the organs designated in Fig. 25b, c,d. The asci are a little wider than in *H. macrosporum*; the most important species difference lies, however, in the size and shape of the spores. While in the case of the spruce parasite the only slightly bent spores nearly attain the length of the ascus, those of the silver fir fungus show mostly a strongly S-shaped bent form and a length which does not attain half the size of the ascus (Fig. 23, 24).

In fully formed asci, almost always the spores do not lie completely next to one another (Fig. 11), but indeed four spores in the upper part and four spores in the lower part of the ascus (Fig. 23b).

Only in individual asci which are retarded in development do all the spores sometimes appear lying together (Fig. 23c).

The emptying of the asci results by a discharge of the spores from an opening in the tip which is only formed by these (Fig. 23a) or by a tearing of the ascus (Fig. 23f). The jelly layer is or is not present in ripe spores; germination follows within 24 hours in May (Fig. 24).

The spermogonia (Fig. 21b) which arise on the upper side of the firmly attached needles in the mountains, or on the upper and lower side of the fallen needles in the flatlands at Neustadt are not essentially different from those organs which were described for *Hypoderma macrosporum*.

Extremely fine hyphae bore through the thick walled cells which lie under the epidermis (Fig. 22a), form a very fine granular stroma (Fig. 22e) from whose surface parallel organs arise at right angles and abstrict spermatia and which force the upper half of the epidermis towards the outside. In spots many of these hyphae grow through the epidermis, in addition to the cuticle, and appear to cut off external spermatia. Since, however, numerous spermatia always stick to the preparation, thus I have not been able to become fully convinced that the spermatia which stick externally to such places of penetration (Fig. 22b) have originated from the hyphae which bored through to the outside.

The spermatia are not essentially different from those organs of the spruce parasite which are illustrated in Fig. 14s.

After lying in water on the microscope slide for a longer time these enlarge themselves, whereby the more rod-like form of these appears more clearly. Germination studies have not been successful up to now, since in a short time extraneous fungus forms develop abundantly and upset the studies.

MODE OF LIFE OF THE PARASITE

The effect of the fungus mycelium on the cell tissue of the needles of *Abies pectinata* is the same as I have described for *Hypoderma macrosporum*. The parenchyma cells shrink together as soon as the hyphae come into contact with them. If the infection occurs in the summer at a time in which no starch is present in the cells, then there also appears in the diseased and killed cells no trace of starch, while if the infection occurs in the month of May, the needles killed from this time up to the winter are completely stuffed with starch.

Apart from the cell contents, the appearances of the disease in the Ore Mountains and in the flatlands (Neustadt) are hardly different.

The most noticeable difference lies only in the almost complete absence of spermogonia on the upper side of the strongly attached needles at Neustadt. Here, individually arising spermogonia occur only on the fallen needles on the underside.

In the Ore Mountains in May, the browning of numerous needles on 2- to 5-year-old shoots begins.

A large part, perhaps the largest part of the brown needles fall. On the needles remaining behind, the spermatia develop on the upper side soon after becoming brown, while on the lower side the formation of perithecia begins usually already in June. In the same year begins, however scarcely, the first formation of asci, which develop only in the late winter and mature in April and May.

At Neustadt the browning does not begin in May, but only in July, the perithecia formation begins in August, maturation occurs in the months of April-June of the next year. The course of development of the disease may be more clear by the following summary, in which the same symbols are used as earlier.

MONTH	ORE MOUNTAINS				NEUSTADT-EBERSWALDE			
	Age of Needles				Age of Needles			
	0/1	1/2	2/3	3/4	0/1	1/2	2/3	3/4
January	.	gr.	gr.	9	.	gr.	gr.	9
February	.	gr.	gr.	9	.	gr.	gr.	9
March	.	gr.	gr.	9	.	gr.	gr.	9
April	.	gr.	gr.	10	.	gr.	gr.	10
May	.	gr.	br.	10	.	gr.	gr.	10
June	gr.	gr.	7	x	gr.	gr.	gr.	10
July	gr.	gr.	7	x	gr.	gr.	br.	x
August	gr.	gr.	7	x	gr.	gr.	7	x
September	gr.	gr.	7	x	gr.	gr.	7	x
October	gr.	gr.	8	x	gr.	gr.	8	x
November	gr.	gr.	8	x	gr.	gr.	8	x
December	gr.	gr.	8	x	gr.	gr.	8	x

For explanation, suffice to say that immediately after the browning of the needles in May or July, the greater part of these fall, and only a comparatively small part of the diseased needles remains attached to the twig. The previous summary applies only to the developmental stages of the parasite on the needles remaining on the twig. Then it is to be considered that not only needles whose shoots are going into their third year can be infected, but that also more frequently than in the case of the fir needle reddening, older needles on up to six-year-old twigs, can be infected. Quite frankly it appears often that in the same year a silver fir which has been healthy up till now is diseased on the needles of the three- to six-year-old shoots, while only extremely rarely do the two-year-old shoots show any diseased needles.

If a tree has already suffered the disease for several years, thus one finds on the three-year-old shoots only newly diseased needles, on the four- to six-year-old shoots newly diseased needles among quite old needles with already emptied perithecia.

The disease results in a defoliation, especially in the lower part of the tree crown, by which the most vigorous stands attain a diseased appearance. This is aggravated still by the fact that the uncast needles on which the perithecia develop remain attached to the twigs for many years before they are completely decomposed.

I have observed the needlecast of fir only in Neustadt-Eberswalde and in the Ore Mountains on 25- to 70-year-old trees; however, the parasite is one of the most widely distributed fungi and the disease in question also must already occur in a damaging degree in other localities.

Explanation of Illustrations

(Plate VI, Fig. 18-25)

Fig. 18. A part of a silver fir twig (underside) attacked by *Hysterium nervisequium* in May (Ore Mountains). Both at the base of the four-year-old shoot, as also at the top of the fifth year's, there occur in part healthy, in part newly diseased and already browned, and in part browned needles with ripe perithecia on the veins.

Fig. 19. Underside of a diseased fir needle in winter. On the leaf vein the perithecia are joined together in an undulating lip-shaped elevation.

Fig. 20. Upper side of a silver fir needle a short time after infection (Ore Mountains). Numerous spermogonia are united into a broad band in the middle of the needle. The margins of this band are mostly strongly thickened and undulating.

Fig. 21. Cross-section through a silver fir needle. On the upper side the ripe spermogonium (b), on the underside the immature perithecium (a).

Fig. 22. A part of the spermogonium enlarged. The extremely tender mycelial threads at (a) penetrate from the interior not only intercellularly, but also, boring through the thick cell walls in the epidermal cells, form in these the stroma (e) from which parallel fine hyphae arise which cut off spermatia on their tips. In spots these hyphae grow through the outerwall of the epidermis and appear to cut off spermatia externally (b).

[111]

Fig. 23. A part of the hymenium from a ripe perithecium. The paraphyses are threadlike, and are not thickened at the tip. Some of these abstrict rod-shaped spermatia (k) on their tip. A jelly layer does not appear to occur on mature paraphyses. The asci contain spores which scarcely reach half the length of ascus and of which four lie in the upper part and four lie in the lower part of the ascus (b). In some rudimentary asci, all the spores lie together (c). Part of the ripe spores possess a jelly layer, part do not. They escape from the asci through an opening in the tip (e) or by a tearing-apart of the ascus (f).

Fig. 24. Germinating spores with and without a jelly layer.

Fig. 25. Immature paraphyses with bubble-like swellings.

Melampsora salicina Lev.

(Plate VI, Fig. 26-28)

The Willow Rust

The culture of the Caspian willow (*Salix acutifolia* Willd., *Salix pruinosa* Wendt., *Salix caspica* Hort.) on sandy soils in the interior of northern Germany in recent years has been managed not always with successful results.

The numerous failures are explained in part by the circumstance that, from ignorance, the Baltic willow (*Salix daphnoides* Vill., *S. pommeranica* Willd., *S. praecox* Hopp.) were cultivated, that unsuitable sites were selected for it, or that the establishment of the willow culture was carried out defectively. In a special paper* I have spoken in detail about this condition, and thereby also described a disease of this willow which is encountered at Neustadt and on other sites in such a damaging degree that the culture of this tree species must be seriously questioned.

The desire to also make known my observations beyond the boundaries of the circle of readers of the mentioned forestry journal, made me decide to publish the same here with the addition of some further results of observations.

The culture of the Caspian willow at Neustadt dates from the spring of 1868. The splendid results of the first years gave the inducement for the establishment of more willow cultures in 1869 and 1870.

The growth of all of these cultures, established in part on better, in part on poorer soils was excellent; the shoots of the 1868 establishment, which were cut off in the spring of 1870 for the first time, already in July of that same year had reached a height of two meters, when on one plant of the 1869 establishment at the beginning of July, 1870, I found the leaves covered with numerous spore pustules of the Uredo form of *Melampsora salicina* Lev. The rust fungus spread from this plant outwards over the neighboring plants, and, in the course of the late summer, over the whole willow establishment.

Several plantings in the meantime have been completely killed; others, so very impaired in growth that their early death can be foreseen.

---

*Die Misserfolge beim Anbau der kaspischen Weide und das Erkranken derselben durch Melampsora salicina Lev. in der Zeitschrift für Forst- und Jagdwesen von B. Danckelmann Vol. 4, pp. 254 forward. 1872.

The parasite which causes the disease belongs to the rust fungi.*

The teliospore form, which winters, is called *Melampsora salicina* Lev. while the Uredo form which belongs to it has been described as *Uredo Epitea* Kze., *Uredo Vitellinae* D.C., *Epitea Salicis, Lecythaea Salicis* D.C., etc.

Tulasne has so thoroughly described and illustrated the parasite in his "Seconde mémoire sur les Uredinées et les Ustilaginées,"** that I can limit myself to summarizing the most important results of this paper.

The Uredo form occurs in summer particularly on the underside, less on the upper side of the leaves, and I can add according to my own observations, also scattered on the bark of the shoots (Fig. 26c) in the shape of smaller and larger pustules (Häufchen). These sharply delimited pustules are colored pale gold-yellow and consist of egg-shaped 16 to 19 micron diameter spores granulated on the outer surface (Fig. 27, 28), before separation borne on longer and shorter slender branches of the fruiting body, and of paraphyses which are thickened club-shaped towards the upper end.

The mycelium spreads intercellularly from the infection point not only in the leaf cell tissues, also often migrates from the tissue of the stipules into the parenchyma of the bark.

Very often I saw at a definite distance from the base of the petiole a circle of spore pustules, after the formation of which a second, somewhat further removed zone of spore pustules arose (Fig. 26c).

The arrangement of these affords the proof that the mycelium spreads itself outward uniformly in the bark tissue from the base of the leaf. The shoots die first in this spot, and later on the whole portion lying above this.

The spores germinate from the pores in a very short time with one to three germ tubes.

If this occurs in moist air on a microscope slide, then these are very thick, similar to those which come forth from the germinating spores of *Peridermium Pini* (Fig. 28).

If, on the contrary, one investigates 48 hours after infection has occurred those spores on the upper surface of willow leaves, which one has brushed with the spores, thus one could recognize that the germ tubes which had formed in the free air are very fine (Fig. 27).

I have been successful without exception in numerous studies in the artificial infection of healthy willow leaves.

According to dry or moist air, the Uredo pustules occur on the eighth day or somewhat later on the leaves infected by brushing with the urediospores.

The second form of the fungus, *Melampsora salicina*, produces survival spores or teliospores. If first appears towards the end of the summer or in the fall in the vicinity of the uredo form.

The cushions of *Melampsora* are in the beginning almost of the same orange-yellow color as the urediospores, but they soon become dirty yellow, brown and finally almost entirely black. When the leaf dies and falls, these cushions have not yet fully developed.

---

*Compare page 63.

**Tulasne: Annales des Scianc. nat., 4. Serie. Bot. Tom. 2.

They gradually enlarge only during the winter, when the leaf is in contact with moist soil, and begin to fructify only at the beginning of the spring.

In the fall and during the winter these cushions, which are still covered by the epidermis of the leaf, consist of palisade-shaped erect cells closely united to one another, the shape of a five- or six-sided prism, whose length reaches 32 to 38 microns, whose width reaches 13 to 16 microns; the spore wall is thick and is colored a light brown.

At the beginning of fructification in the spring the epidermis has become thinner and sometimes appears to be reduced to the cuticle, which is almost fused together with the upper ends of the palisade-shaped fungus spots. Fructification proceeds as follows: usually at the end, more rarely at the base of the prismatic cells, cylindrical tubes (promycelium) arise, which usually send out four side branches, each of which on their end bears a round, light gold-yellow sporidium of about 10 microns diameter.

Simultaneously with the formation of the sporidium, the promycelium divides itself into as many cells of unequal size. The sporidia are easy to germinate, and give rise to the disease again.

The two described forms of the parasite frequently occur next to one another. However, likewise often one sees the Uredo form or the *Melampsora* cushion developed almost alone and encrusting the upper surface of the leaves. On dead leaves only the second form develops, which grows in the first stage as a parasite, in the second stage as a saprophyte.

The question whether the *Melampsoras* which occur on different species of willow belong to the same species, or whether there are specific differences has recently been decided more in favor of the first opinion.

However, I hold this question as still not decided and will note only that while the infection of leaves of *Salix acutifolia* succeeds without exception by the urediospores of the fungus, repeated infection studies of *Salix daphnoides*, *S. purpurea, S. nigricans, S. silesiaca, S. triandra, S. viminalis, S. pentandra,* and *S. cinerea* with the urediospores from diseased leaves of *Salix acutifolia* came to nothing.

I intend to continue these investigations and to undertake infection of Caspian willow with spores of *Melampsora* from leaves of other species of willow.

As concerns the progress of the disease and its spread in general, those plants which are attacked first show in the beginning only a few yellow flecks; soon the number increases and if a larger number of leaves is attacked in the earliest stages of development when the leaves are still unfolding, such numerous flecks occur that within a few days the leaves become yellow, finally black, roll up and fall off (Fig. 26b).

Since, as I have reported, the shoots also are often attacked by fungus mycelium, many of these die at the tip.

If the disease occurs early in the year, as was the case in 1870 where it appeared already at the beginning of July, then the plants, deprived in a short time of their leaves and tips' ends, seek to releaf themselves anew by the development of lateral shoots. However, these also were soon attacked and killed and thus not only the growth of the plants ceased quite early, but these also were not in a condition to form reserve materials for the next year.

The killed twig ends give proof of the occurrence of the disease in the winter and for several years.

The course of the disease at Neustadt was as follows:

While in 1868 and 1869 no trace of a parasite had been noticed, it appeared first at the beginning of July, 1870, on the shoots of a stem of the 1869 culture planting. From these the rust fungus spread itself out within 14 days up to the middle of July on three two-year-old and one one-year-old willow plantings which lay in the vicinity. Separated from this planting by an about five meter high thick spruce hedge, the 1868 willow bed showed the first traces of the disease only about the middle of August, at a time when the first beds were already fully yellow and rather defoliated.

One must assume that the thick spruce hedge had hindered the dissemination of the rust spores for a long time.

Three willow plantings from 1870, which were lying somewhat removed from the other willow plantings, remained healthy for a long time until they were also infected in September.

At the end of the first year in which the disease occurred, there was a planting, established in the spring of the same year and attacked by the rust in the first half of July, which was completely killed; three other plantings of the same age, however, were attacked only in September, and showed only killed twig ends. All older plantings remained in the stage of development in which they were at the beginning of the disease. Several stems were withering only on the first diseased spot. In 1871 the growth of willow in all plantings was more sparse in comparison to the previous year, scarcely half as large as earlier.

In the first days of August, the plantings attacked last in 1870 were infected first and to the extent that they were completely killed before the beginning of winter.

The other beds were first infected towards the end of August; in spite of this, two were almost completely killed by fall.

In 1872 at the first of June the parasite occurred already in a sample of the 1869 planting; in this vicinity by the 18th of June all plants had been attacked by the disease.

On the ninth of July all willow plantings which lay in the vicinity of this spot were nearly defoliated.

In 1873 the disease began at the end of June and already by the end of July had spread over the still existing three willow plantings. Two-thirds of all plantings were destroyed in the course of a few years; the rest so very much set back in growth that the survival of these can scarcely be expected.

The experiences until now indicate that new plantings, if they become attacked by the rust fungus in June or July, exhaust themselves completely in the first year and die.

For a time the plants seek to produce new leaves and shoots at the expense of the building materials stored in them. However, since these are always soon killed again, consequently no new building materials can be assimilated, thus already in the late summer and fall complete exhaustion and death occur.

If new plantings are first attacked in the fall, also in August or September, then the assimilative ability of the plants will be broken in such a manner that a collection of reserve materials can no longer occur and therefore reduced growth occurs in the next year; however death does not occur yet.

Older, two- to four-year-old plants are more resistant; nevertheless, they decline very much in growth not only due to the fact that with the occurrence of disease, growth for that year ceases, but also due to the fact that the deficiency of reserve materials brings about reduced growth in the next year.

After several years most stems die. As usual, this disease also is hastened by wetness of the air, since the germination of the rust spores which stick to the leaves is thereby hastened.

In cases where the first occurrence of the parasite is noticed in time, cut out and ruthlessly burn all attacked plantings in order to limit as much as possible the spread of the disease. Complete elimination of it will not be attained because of the easy contamination by the spores.

In order to hinder or at least limit the carrying over of the disease from one year to the following, it is recommended to experimentally rake together and burn in fall or winter the leaves which possess the *Melampsora* cushions.

Since the germination of the survival spores occurs only in the spring on the killed leaves, the pustules are covered by the epidermis, the spores also prior to germination are not disseminated on the soil or in the air, thus this permits the means of control to be brought into practice perhaps on a large scale.

Here at Neustadt I have not until now wished to perform research on the extermination of the parasite since I would thereby deprive myself perhaps of a very welcome teaching aide for pathological lectures and demonstrations.

Explanation of Illustrations

(Plate VI, Fig. 26-28)

Fig. 26. Twig of *Salix acutifolia* attacked by *Melampsora salicina*. a. A still green leaf with numerous fungus cushions. b. Leaves which have already become black-flecked and rolled together. c. A fungus cushion breaking through the epidermis of the stem.

Fig. 27. Spores of the Uredo form of *Melampsora salicina* germinated in free air.

Fig. 28. The same, germinated in a moist room.

Postscript

To *Agaricus (Armillaria) melleus* L.

Before the end of this printing I am in position to describe the form of mycelium which issues directly from the spores of *Agaricus melleus.*

The spores from the caps of the fruit bodies in the fall form a white powder which often is detectable on the moss lying under it, on leaves, twigs, etc., often still a year afterwards.

Investigation of this white coating in September of the following year at the time when the new bodies were forming, showed that in place of the spores, only a small number of which could still be detected in an almost unchanged position, a thick texture of hyphae had appeared, which by their clamp connections and granulated outer surface fully agreed with those hyphae illustrated in Plate I, Fig. 8, and Plate II, Fig. 9.

Here and there these united into finer and thicker hyphal strands which have been described as outgrowths of *Rhizomorpha*, Plate I, Fig. 5c, 6, 7, 8, 9.

The supposition that this mycelium had arisen from the previous years spores of *Agaricus melleus* was thereby significantly strengthened in that also the earth in the vicinity of the rootstock also was abundantly grown through by the same mycelium and numerous small fruiting bodies of *Agaricus melleus*, of which, however, most were vestigial and the largest attained the size illustrated in Plate II, Fig. 15, were growing out of the soil, without being attached to the developed rhizomorph strand.

Unfortunately conditions did not allow the investigation of the formation of fruit bodies from this mycelium.

Just as thicker and thinner non-medullated hyphal strands branch out (Plate I, Fig. 5c) from the normal rhizomorphs, the latter are able to form into rhizomorph strands under favorable conditions, and I believe hereby that the still unanswered question of how rhizomorphs can arise from the spores of *Agaricus* to have been answered with the greatest probability.

As regards the time of fruit body formation, I found this year already on the 24th of September on a trip in Solling, and a few days later in the Harz Mountains some spruces on which the stage of development illustrated in Plate II, Fig. 5 was detectable, while on the mostly killed plants the fructifications stood at the stage illustrated in Plate II, Fig. 4 and 6.

On the 15th of October near Neustadt-Eberswalde only a few fruit bodies were found, on which the cap was still not fully developed. Weather in 1873 has hastened the fructification about 14 days earlier than in 1872.

According to a report from Forest Supervisor Danckelman, he has seen very numerous fruiting bodies of *Agaricus melleus* growing out from the bark cracks to a height of about two meters on a Scots pine about 15 cm in diameter, which in any case is rare, since the bark of Scots pine is so thick that it does not easily crack.

Rhizomorpha subcorticalis, as I have already reported, easily ascends older Scots pine under the bark up to a height of two or three meters.

Morphologically interesting is a fruiting body which I found a short time ago, on the upper side of the cap of which a somewhat hazel-nut-sized outgrowth occurred. This was split open in the middle and the funnel-shaped cavity showed the star-shaped lamella running out to the split margin.

In respect to the life cycle of the parasite, I report only that also spruce, like Scots pine, does not remain exempt from the disease at greater ages. In the Holzmindender district on the Solling I found an approximately 45-year-old very vigorously growing spruce about 30 cm diameter killed by the parasite. The thick mycelial felts had grown upwards under the bark several meters high.

According to a statement from Forest Supervisor Dürking, in the vicinity of this tree a thicket of vigorous spruce were killed a few years before without recognizable cause and had been felled. A supposition is obvious that these spruce also had been killed by the parasite.

To *Trametes radiciperda* R. Hrtg.

Until now I have found only young samples of this *Trametes* which I have illustrated in Plate III, Fig. 20-23. An at least five-year-old specimen found by me at the foot of an old Scots pine killed by this fungus showed the characters of the fruit bodies more clearly.

The shape of the fruit body can best be brought to mind if one thinks of a bracket-formed fruit body of *Trametes Pini* illustrated in Plate III, Fig. 4, the lower half removed by a cut parallel with the upper sterile horizontal surface.

There arose in this fashion a second smaller sterile surface oriented towards the bottom while the porous sloping hymenial surface on both sides is nearly as broad as in the middle. The snow-white, swollen, pore-free margins of the hymenial surface are very irregularly furrowed and folded, similar to the margin illustrated in Fig. 22. The sterile horizontal upper surface is irregularly humped, however, shows the narrow swollen margins formed at the end of the growth periods above the common upper surface. With the exception of the snow-white margin, the zone formed last year is rust colored and covered with fine velvety hair (Plate III, Fig. 27).

The older portion of the surface is, on the contrary, darker brown and grown over by a fine moss. In a perpendicular section the annual zones of growth can be distinguished clearly from one another, in that a narrow uniform layer occurs on the boundary, which forms at the beginning of each growth period and in which the canals arise anew.

[119]

The older inner canals were filled with a snow-white filling material while the walls of the canals show yellowish-white coloration. The substance is corky, somewhat shriveled together by drying, and is very hard. In a fresh condition this has a pleasant smell similar to that of *Boletus edulis*. The shape of the pore opening is similar to that of *Trametes pini*. The length of the canals reaches about 5 mm each year. The width of the fruit bodies in hand measures 8 cm from one side to the other. The distance from the upper to the lower sterile surface is about 2.5 to 3 cm, the width of the sloping hymenial surface in the middle is about 4 cm. The distance of the upper swollen margin from the stem that is the attachment point in the middle is about 5 cm.

Pulling up younger plants and tearing out the stems of older plants killed by the fungus is advisable in order to prevent the spreading of the fungus by the spores produced by the fruit bodies which still grow for a few years on dead plants.

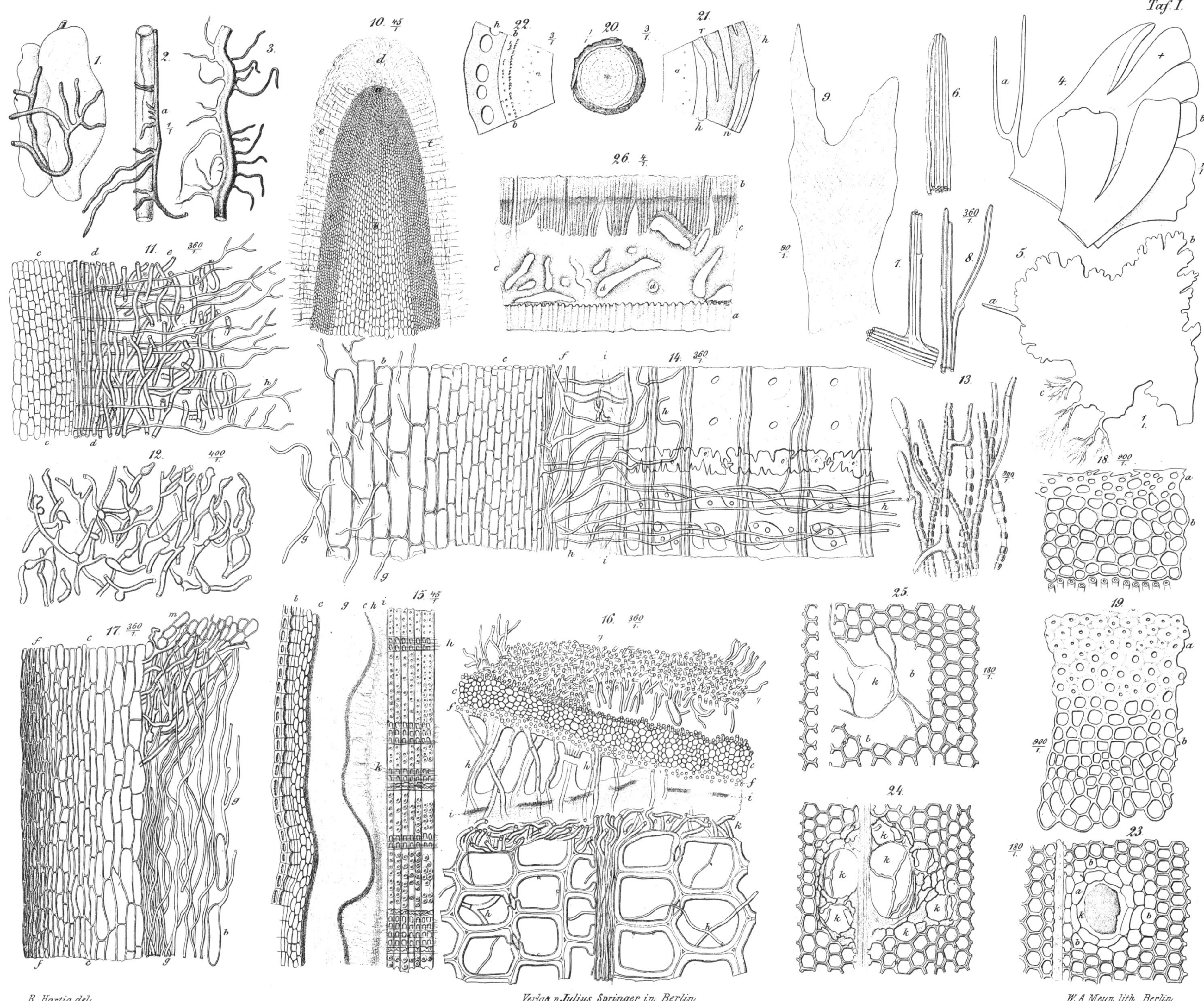

Taf. I.
R. Hartig del.
Verlag v. Julius Springer in Berlin.
W. A. Meyn lith. Berlin.

R. Hartig del.
Verlag von Julius Springer in Berlin.
Lith. v. W. A. Meyn. Berlin.

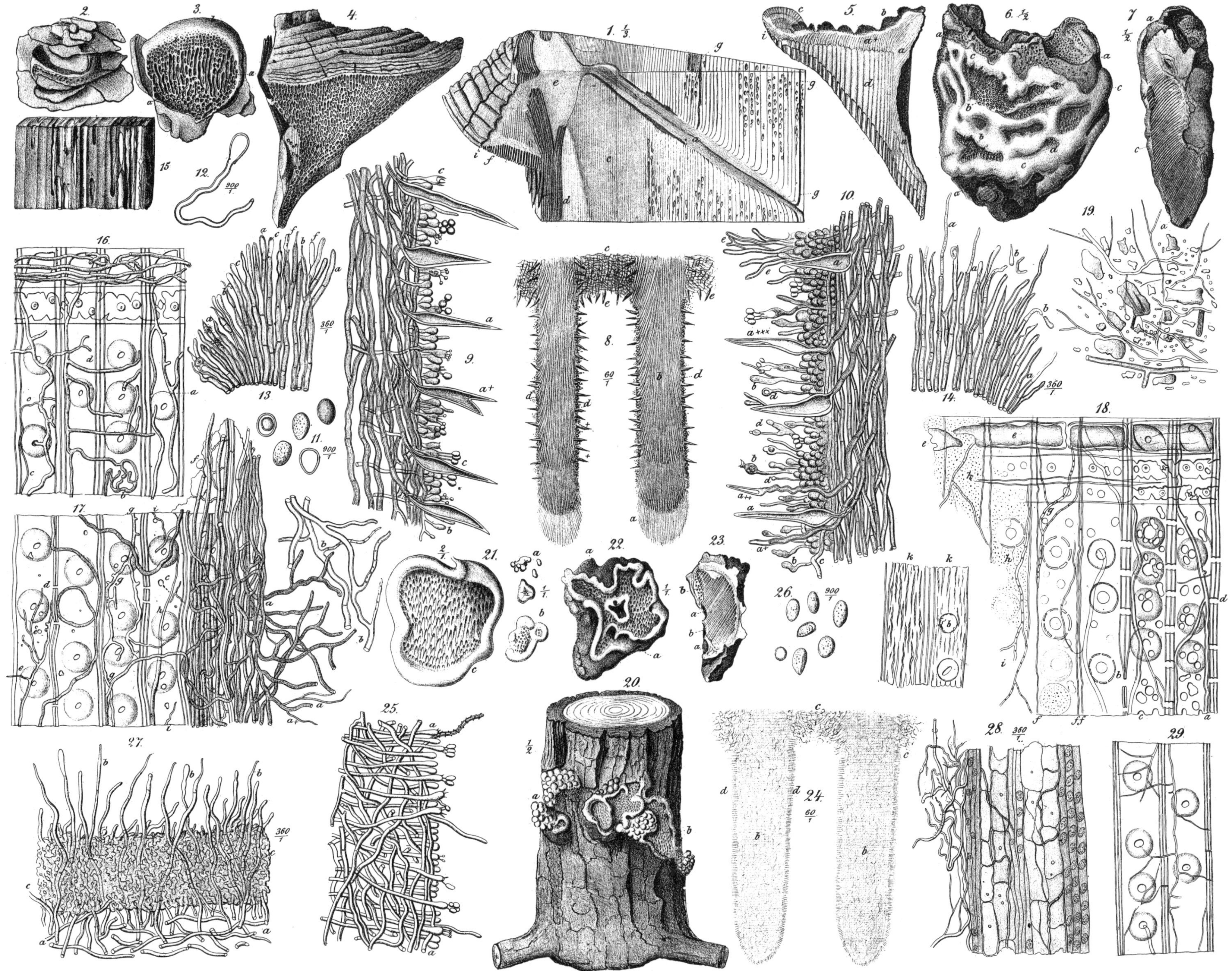

Taf. III
R. Hartig del.
Verlag v. Julius Springer in Berlin
Lith. v. W. A. Meyn, Berlin

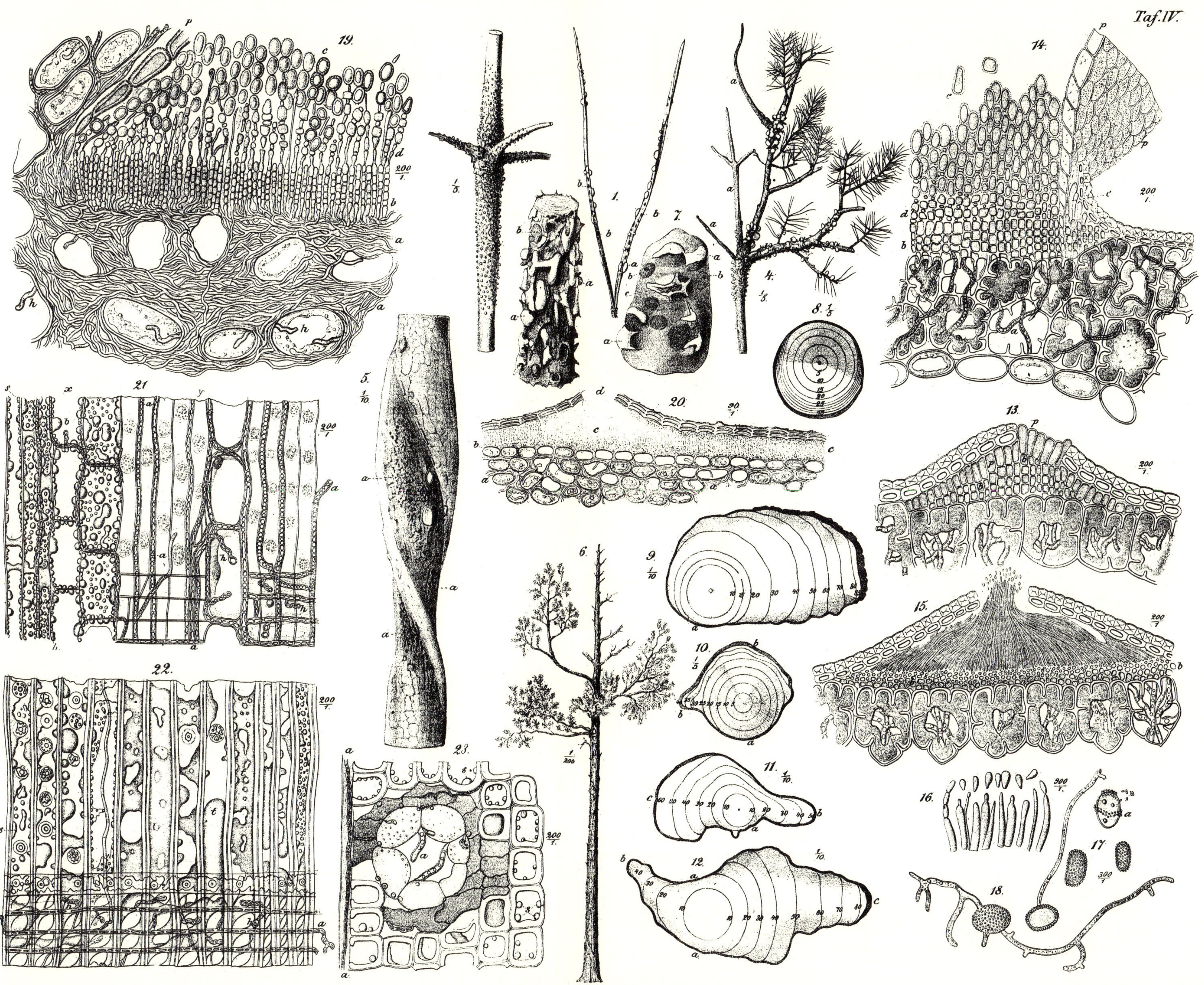

Taf. IV.
R. Hartig del.
Verlag von Julius Springer in Berlin.
Lith. v. W. A. Meyn in Berlin.

Taf. V.
R. Hartig del.
Verlag v. Julius Springer in Berlin.
Lith. v. W. A. Meyn, Berlin.

Taf. VI.
R. Hartig del.
Verlag v. Julius Springer. Berlin
W.A.Meyn lith Berlin.